uneasy money

uneasy

BY THE AUTHOR OF *plowman's folly*

"Here to the homeless child of want
My door is open still."
—GOLDSMITH

money

EDWARD H. FAULKNER

NORMAN : UNIVERSITY OF
OKLAHOMA PRESS : MCMXLVI

To My Wife

THE CHAPTERS

uneasy money

NEEDED: A NEW VIEWPOINT

T HE AMERICAN PEOPLE—some six or seven generations of them—have enjoyed a period of economic
circumstances which may correctly be called an era
of easy money. I haven't been rich; nor have you perhaps.
But we judge ourselves, necessarily, by the only standards
of wealth we have experienced, those of America. Thus,
as we view our own finances we have not been wealthy and
know of no reason why anybody should think we are. Indeed, so many others we know have so much more money
than we have that we often wonder whether we ought not
think of ourselves as among the poor. Yet, as the rest of
the world views America, our country is the one country
where everybody can buy whatever he likes.

We think differently, and with good reason. We must
think and act within the economic framework in which
our lives actually develop. If we should suddenly accept
the rosy-tinted appraisal which our envious world neighbors have adopted for us, and should spend money in ac-
cordance with that appraisal, we would soon be on our
uppers financially—and in a country where money is the
economic blood of life.

Until recently it has not been necessary for Americans to worry about what our world neighbors thought. We were comfortably near self-sufficiency as a nation. We sold our products to every other nation. We could dismiss with a shrug all disapproval of American ways. Our financial position was such that we could—in a world at peace—ignore the taunts of foreigners who called us "money-mad."

We find ourselves today in quite different position—not that we are any the less self-sufficient economically, but rather that our historic isolationism has been abandoned for membership in the United Nations, with all of the responsibilities that such membership entails. What others think of us now becomes very important, for, from our characteristic position of aloofness as a nation, we are being thrust into world co-operation, economically as well as politically. It is a role we cannot escape, nor should we if we could.

It is one thing, however, to join a world organization and quite another to adjust ourselves as a whole people to the new role that this action assigns to us. All our previous experience and training, our home life, and our political point of view have contributed to the development of ideas antithetical to those now required. We have been taught, almost as a religion, to mind our own business. World affairs were not our business. Whatever was American was right, and our concerns ended there. And why not? Similar teachings prevailed in most other countries. Peoples elsewhere were taught to be intense nationalists. America, more perhaps than any of the rest, however, was a dream country, and the pride that has been engendered here has outweighed our thought of other patriotisms existing elsewhere. At the same time, it has blinded us to certain needs and hopes in other lands, the fulfillment of which may become partly our responsibility in the postwar world.

To grasp fully the points of view of diverse world peoples is, of course, an impossible task. The elephant could scarcely achieve the full life views of the mouse that shares his quarters and nibbles his food. Yet both have equal right to full enjoyment of life. The dissimilarities between our own normal lives and those of Eskimos, Pygmies, South Sea Islanders, and Japanese are almost as great in some ways as the differences between the life of an elephant and that of the mouse. We need not expect to comprehend easily the peculiarities of our world neighbors. We must keep in mind that whatever in their lives seems strange to us, even inexplicable, is perfectly normal to them; also, many of our most sacred customs seem equally baffling to them. We don't eat grubs; but as consumers of shrimp we shouldn't quibble about that. We don't eat snakes or insects. Hindus, on the other hand, don't eat cows, and may resent our doing so. Indeed, many peoples of the world eat no meat, thus avoiding a nutritive waste, since pigs, cattle, and sheep must maintain their bodies while adding weight, and part of the food they eat must be used up in body maintenance. That fact, rather than vegetarian principles, explains the vegetarian diet of many peoples. Land limitations dictate diet as often as do preferences. We in this country have known no such bar to food abundance; so we will do well to substitute a studious interest in other peoples in place of our usual patronizing attitude.

Until we have achieved an understanding of the historical and social backgrounds which have conspired to make people different from each other, we here are ill-qualified to plan the postwar future of the peoples of other countries; it is even doubtful if we are qualified to plan our own postwar future in a world where we must manage to live at peace with these other peoples whom we scarcely understand. All the planning we do for ourselves or for our world neighbors must necessarily be within the thought frame of our understanding; and, until we have become conscientiously curious about what animates the thinking of the many peoples of the world, we cannot have the cosmopolitan viewpoint from which to wisely help them plan their future in a world that hopes to co-operate for permanent peace.

Many an American will take a dim view of the merest suggestion that the American viewpoint is other than cosmopolitan. This country has led the world in so many matters that it is easy to assume that Americans would understand world peoples better, perhaps, than would the people of any other country. The truth is that we have been so busy leading the world in the development of new ideas and the machines to implement those ideas that we haven't had time to inquire into the lives of the people all over the world who have been our customers. We have known our Churchills, our Chiang Kai Sheks, and our Francos; we have known little of the cockneys, the coolies, and the Spanish peasants. In short, we have learned little of the ways of the non-American world; ways that are as significant to the peoples concerned as they are incomprehensible to us. If we are wise, we will lose no time in making the acquaintance of these everyday details which for many years the *National Geographic* has been glamorizing for its readers.

For much such information we need not go abroad. At least, we can get a good introduction to the real people of the United States by venturing off the paved highways and rambling easily through the rural neighborhoods of our country. Meet the "hillbillies" on the home grounds. Realize the fact—and it is a fact—that there is something far more significant than the merely quaint about these students of the university of hard knocks. Generations of wresting a living from land that has been classed as "submarginal" or worse has taught them many lessons that are not to be found in books. They are prepared to continue indefinitely fighting through to another year's food supply, despite what would appear to many of us to be insuperable odds. How do I know? That is a fair question. The answer is that I grew up with such people. For four years I served the people of my home county as their first county agent. During those four years I was fed and lodged about four nights of each week under the roof of some back-country farmer. And all the hospitality that was extended to me was spontaneous and generous. Few of even the poorest farmers would consider accepting pay; indeed, to many the offer of pay for the substantial favor of food and shelter, as well as care for my horse, was almost an insult. Those people hungered so much for news of the county seat, or of farm conditions elsewhere, that they treasured the opportunity to have as a guest someone who could satisfy this craving.

Traveling by horse was necessary, because in the 442 square miles of that county there were but thirty miles of pavement outside the streets of the small towns. If you doubt the existence of poverty in America (as of World War I days), consider the fact that I spent one night in a house where the family consisted of man and wife, mother-in-law, and several children; and that the one pan they owned served for dressing the breakfast chicken, for making the biscuits, and as a wash basin for the guest. Often I have heard cynics say "Yes, the Kentucky mountaineer is hospitable; he'll make you welcome to the best he has; but he has nothing to offer." Literally, the statement is in some instances true, but how false spiritually! The good people with whom I lodged that night gave the little they had without apology, made me feel welcome, and invited me to stop in again whenever I was close by at nightfall. What more could they have done?

Perhaps that is a very extreme instance for any part of the United States today. I have no way to know. We may be sure that it is not extreme for much of the rest of the world. If you travel far enough you will find humans living in every conceivable kind of structure, and enduring every degree of insanitary condition. These are conditions that our leaders have determined to stamp out. We are determined as a people to initiate the rest of the world to bathtubs and soap—to every American "advantage," in fact. The task we have set for ourselves is to raise the living standards of a billion or so of the world's inhabitants—by selling to them the American made products that have enabled us to climb upward from conditions similar to those we find elsewhere in the world.

A large percentage of these billion prospective customers for American goods seldom see a coin of any kind. As they have lived, and as they continue to live for the most part, the majority grow part or all of their own food; and in many places the fiber for clothing is also produced by animals or plants, and the clothing actually made at home. In other words, among such people there is little trading, and, therefore, little use of money. Despite the virtual absence of cash, these people live obviously happy lives, something we in America have often forgotten how to do. Indeed, the development in America of a situation in which most of our people depend upon buying their food, and all of us buy our shoes and clothing, creates an extremely sensitive economic balance. Few of us, even with the abundance of gadgets of civilization we possess, can live the carefree lives that primitive peoples enjoy. They have the advantage that their food supply can't be interrupted save by nature, and that seldom. They grow it themselves.

Yet we are agreed these people must be uplifted. And we have made a beginning by furnishing machinery and engineering skill to several of our backward neighbor countries. This will help many moneyless people to acquire the necessary cash with which to buy the things we have to sell. But there are limits to the extent of improvement we can make in this way. Because our goods will cost them too much, they will buy sparingly. Hence we must supplement this mechanization of foreign countries by *drastic reductions in the selling prices* here and abroad *of all American products.*

The bare suggestion that American goods must sell at lower prices will startle readers who have followed the history of economic development in the United States. They will see both labor and management pouncing upon the suggestion from the industrial side; and, of course, everybody knows that farmers must have higher prices—not lower, certainly—if they are to be active customers for our own industrial products. This pattern of thought is so valid for every student of American economic history that most readers will be puzzled to know how my theory will square with reality.

The lower-price-scale suggestion, however, is both valid for us as a nation and absolutely essential as a factor in successful future world trade. We can probably increase our volume of trade slightly by measures designed to help the customer find the money through employment in his own country; but if our prices are at a level to yield the necessary profit to our own people, the trade volume will necessarily be very limited. If the government subsidizes such trading as a means of reducing the selling price in the foreign market, we shall be like the apple salesman who bought his stock at five cents and retailed it at six for a quarter. The more export business of that kind we have the worse off we shall become.

Little or no thought has been given to the possibility of slashing production costs of everything grown or manufactured in this country. There are two very good reasons for this apparent negligence: (1) for several generations we have lived with the philosophy that high wages and high prices are essential to prosperity in this country; and (2) the difficulty of accomplishing any great decrease in production costs is generally accepted. For farmers, particularly, cutting costs is considered impossible.

Despite this apparently impossible economic position, almost every economist who discusses postwar relationships announces premises from which the most obvious conclusion to be drawn is that our general scales of costs must be reduced. Yet none, as far as I am aware, has ventured into that uncharted region. Instead, fearing to tread such a supposed quagmire, most economists conclude that the government will have to solve the dilemma by such subterfuges as can be agreed upon.

In later chapters I show how value levels can be reduced by starting the cost cutting with agriculture. The basis for lower farm production costs will be questioned, of course; but there is both a theoretical and a practical basis for such argument. Theoretically, American farmers are either as capable of as good a production job as, for instance, Chinese farmers, or they are not. I do not believe anybody will argue against the ability of our farmers to equal the production of the farmers of any other country in the world; hence, if that is not arguable, then, by inference, American farmers can grow as much per acre as can any Chinese farmer. And, since our farmers will use machinery and grow several times as much per man, in addition to equaling the Chinese production per acre, they will produce these crop yields at a lower cost per unit than they do now—possibly lower even than the cost of production in China.

It is unnecessary to add what truthfully might be said: that this greatly increased production per acre and per man can be accomplished without any soil aids other than those which the land itself can produce under proper management. In other words, instead of emulating the odorous practices of the Chinese farmer who uses human wastes, the American farmer will grow his own organic matter, work it into the soil mechanically instead of laboriously composting it, and come up with just as high production per acre.

So much for the theoretical aspects; now for the practical reasons for assuming that American farmers can grow cheaper products than they have produced before.

In each of the past several seasons American farmers already have produced a greater total from the land and feed lots than was produced in the next preceding year. In many instances there have been surprising and gratifying increases in production per acre. These increases have generally been credited to unusually favorable weather conditions; but it is known that in many places, even though the weather was decidedly unfavorable, there have been unexpected yields—sometimes increases over former years. These facts indicate that farmers are now benefiting from following the better methods that have been taught in the past decade by the Soil Conservation Service and other agencies; and, since such effects are necessarily cumulative, it is to be expected that the future will witness progressive increases in production.

The last decade of instruction farmers have received has not been wasted, in my opinion. Throughout all or most of that period farmers have received advice from the Soil Conservation Service, the Agricultural Extension Service, the Farm Security Administration, the Agricultural Adjustment Administration, and many other less prominent agencies, both public and private. Unless all this effort has been fruitless, such results as I have outlined are normally to be expected. Better practices in managing the soil can even mitigate the effects of bad weather; and I believe that has happened in many instances.

Among the necessary results of such an improved situation must be decreased costs per unit of growing farm crops. And, when sizable cuts in farm production costs have been made, the consequent lower prices must bring down the cost of living for all of us. This, in turn, opens the way for a round of cost reductions throughout industry, made possible by the lower costs of living.

Already the trend is under way, largely unrecognized. That fact makes it highly important that certain changes be made in the established system of crop rotation, which, I believe, has been responsible in part for some of our serious crop surpluses. The complications of this phase of the matter are too great to be introduced here, but I suggest that the reader follow through the text for more detailed discussion of the whole matter. The situation we shall be in if we drift helplessly into a whirlpool of chaotic prices justifies the title *Uneasy Money*. And the man who is entirely disconnected from the land is the one who will suffer most.

2

RAGS TO RICHES–AND BACK

T HE ROUND TRIP from rags to riches has been traveled by many an ambitious American—by some of them several times. That an entire country could travel that same trail may not be so familiar a thought, and the suggestion that our country is even now doing that very thing may create puzzlement in orthodox circles. There is good reason for thinking, however, that such is our course, hence the need for examining critically some of the reasoning upon which such an unusual judgment is based.

The discovery of America was an accident, the import of which we have been too busy to appraise fully. Columbus thought he had found the land from which Europe for generations had been receiving the spices with which its kings and princes added pleasant flavors to their food. It never occurred to him then, nor to anybody else for several generations, that he had found, instead, the one remaining big area of untouched soil from which an increasingly hungry world could feed itself.

In those days Europe still could feed its people comfortably from soil that had been fought over by the Roman legions and had been partially peopled by them. The growing of food was still easy. It was still possible for a population, thinned out by the Black Death of a century and a half earlier, to subsist upon the available land. Indeed, the Age of Chivalry—in which feasting, at least for the favored classes, was an important ceremony—was still ahead. Europe did not immediately need America's products; but the time would come.

Meanwhile, the peopling of this continent developed haltingly through the next few centuries. The absence of anything that could be suspected as planning was one of the conspicuous characteristics of this haphazard development. Certainly none of the first settlers came here intending to found the greatest industrial nation the world has ever seen. Many diverse motives brought people here. Some came from curiosity; some to hunt gold; some to trade with the Indians; some to find refuge from persecution; and not a few were unfortunates, dumped here to relieve jails crowded with people who couldn't pay their debts. From such a drab beginning this country got its start; but by the early part of the eighteenth century the timorous attitude with which many early venturers came disappeared, and settlers began to come in bold migrations.

Since this continent was peopled very sparsely by Indians who virtually lived off the wilderness, visiting Europeans soon discovered the opportunity it presented for exploitation. As the news spread, the incoming groups increased by leaps and bounds, making it seem wise, eventually, to restrict immigration. This seems an odd neces-

sity, since every newcomer was another mouth to feed—a
home market for the surplus food we grew so easily. Also,
every new family needed some kind of house—cabin, sod
house, cottage; eventually perhaps a mansion. Then there
was furniture: at first rough chairs, a table, and a hemp-
rope bed in a corner; but for most of us the time came when
it included carpets, draperies, and pictures on the walls.
And, of course, clothes. The coonskin cap of the pioneer
was no more permanent than the cabin in the clearing.
Both gave place in time to luxuries which, in this country
as in no other, were shared by a great proportion of the
population.

Such restraints as we placed in the way of immigration
grew out of fear that people who had been ill-paid in Europe would be quite willing to work here for wages that
would seem high to them, but would be disastrously low
for workers already accustomed to spending a higher income. This immigration, which in its early stages brought
us some of Europe's best brains as well as its brawniest
and most venturesome yeomen, was essential at that time
to supply the population increase we must have for the
country's development. Later, when the ocean lines began bringing a million one-way passengers a year to this
country, it seemed as if all Europe was moving here as
fast as relatives already arrived could send back enough
money for immigrant passage. These newcomers were all
used to earning little and spending little; when they got
jobs here they earned much, spent little, and sent the balance across "the pond" to help their parents, brothers,
and cousins to get over here, too. The prospect that an

oversupply of labor would result from this ever-increasing number of workers seemed so definite a threat to our established standard of living that our government sorrowfully shut off this wild influx.

The general scheme of the prosperity we enjoyed for some two centuries began to develop as soon as enough of the forest had been cleared to permit the production of farm crops in excess of what our own population could consume. In the earliest stages of our world trade we shipped out furs and little else; then we added farm products; later lumber; and considerably later, coal. Of course, there were other exports, but these were among the most prominent before we developed our own industries. It is easy to understand that, with the rapid settlement of the country, hurried along by the building of railroads into the plains country, the volume of our exports mushroomed to great proportions in a few decades. And, while we gave no thought then to such effects, it is easy to realize that, as we supplied the world with cheap cotton grown with slave labor on our fresh land, we cut into the established trade of Egypt and India; similarly, machine-grown wheat from the "inexhaustible" black soils of our Great Plains made important inroads into the grain trade of the world. Being food, wheat did not so completely supplant its competition as did cotton, of course. Wheat for domestic consumption continued to be grown in most places where it had always been grown. Cotton, being produced for manufacture, lost out completely in most areas as soon as low-cost American cotton entered the field in sufficient volume. For years it was virtually a world monopoly for us.

Our country had many advantages. Land was both new and plentiful. Every farmer could clear and own far more land than he could manage by hand. The easy money to be made in the world food and cotton trade dictated the universal use of machines in farming. So long as the land remained naturally productive, machine farming gave us still further advantages in trade. Fortunately for the countries whose markets we appropriated, we needed in this country many things we were not equipped to produce. Such manufacturing as we did was chiefly farm implements. Other necessities and luxuries we were glad to buy abroad. We had a special advantage over our competitors in selling farm products in that we omitted to charge into our costs an item for plant food replacement. The worn land of others could not compete on even terms with our fresh soils because of that omission. Through such special advantages we managed to accumulate here in a couple of centuries the greatest pool of negotiable wealth ever brought together in a single country. It was no accident; nor was it the result of genius; it consisted simply of resource liquidation.

We could easily and mistakenly consider this wealth the crowning advantage of all. In some ways it undoubtedly was just that. It made possible here the greatest common participation in high living standards that has been known anywhere. It made possible the world's greatest use of fine mechanisms and their application to the greatest variety of uses. Too, it has made us an important factor in the termination of two world conflicts. There have been advantages too numerous to mention; but the possession of so much money has not been entirely advantageous. There have been disadvantages too; unrecognized, but disadvantages nevertheless. Of these but a few can be named here.

Many of the handicaps of the possession of money are necessarily psychological. Obviously this is true, since one easily finds ways to put money to good use. Too easily, also, money finds inappropriate uses. Like most other useful things, there can be too much of it. Fire and water, for example, are indispensable; but we can have too much of either. The disadvantages of too much fire or too much water are so much more apparent that we easily recognize them; the handicaps of too much money, being chiefly in the creation of incorrect mental attitudes, are less evident.

One of the chief psychological disadvantages of our position as the richest people in the world has been our failure to recognize how we came by that wealth—to credit its accumulation, therefore, to American ingenuity, and to reason from that point of view that other people the world over could likewise be equally wealthy if they but had the genius that has been displayed by American businessmen. We have, on that account, always been somewhat contemptuous of the people of other nations. On this account it was easy for our statesmen in the nineties to convince us as a people that Orientals, reputed to require but seven cents a day for maintenance, were not suitable to be admitted to citizenship here. Our industries would be ruined by such cheap labor, they argued. Dozens of other ways in which we have stood in our own light, due to our wealth complex, might be cited.

Money, too, has been substituted for labor; and in ways we are tardily recognizing it has done yeoman duty for thought itself. Nobody would question the suggestion that money has been substituted for energy, labor, or time. We use electricity wherever we can to save work. We hire people to do for us almost everything we are not compelled personally to attend to. We speed our travel, our communications, our living in general by the use of money.

The ways in which we have used money as a substitute for thought are a little less evident. The way I would think of first, of course, is in our handling of our soils. We have thought so easily in terms of banking that we substituted analogies from banking to explain our soil troubles, disregarding the fact that plant growth is in the field of biology, not banking. We have confused the soil as a self-renewing resource with a bank account which, obviously, is not self-renewing. We have created so much confusion in the minds of farmers by such pseudo-thinking that they no longer remember to take their cues from nature rather than from the bank-account analogy. The great American blasphemy is our implied accusation that the Creator left us here in a world of continual and necessary warfare against nature.

Another way in which we lean on our money is our persistent promotion of research into problems for which we already have the answers, either in previous uncorrelated research or in recognizable form in nature itself. Tons of paper have been wasted in research, the results of which we have failed to profit from. It may be said with fair accuracy that, unless somebody who can profit from a bit of research takes in hand the results and commercializes them (vitamins, for instance), we pay little or no heed to research findings of even the most vital character. If you will take the trouble to list the supposed advances agriculture has made, you will find that almost every item science has recommended either has been commercialized or is not widely used. Commercialization explains to some degree the extent to which farmers use fertilizers, lime, pest-control items, legume inoculating materials, and other "needs" of farmers. If nobody could make a profit from the manufacture and distribution of these things, the research findings that seem to justify their use probably would be gathering dust along with others—possibly even more valuable—which have not found a commercial outlet.

Oddly enough, it is easy to point out in nature demonstrations of facts of value to farmers that require no verification by systematic research. Yet, though these demonstrations may be very evident and may prove their value with emphasis, their example is passed over as of no consequence. Apparently we prefer test-tube judgment to natural evidence. Landscape proofs are free; those from the laboratory cost money; having the money, we seem to prefer to spend it.

Such a point of view doubtless will be questioned. For evidence as to its validity, we need but consider agricultural practices in countries where there is less money. When we see that farmers in those countries have neither science nor money to lean upon, as our farmers have, we can but wonder what miracles save them from the goblins that would get our farmers in such a poverty-stricken situation.

Our farmers probably learned to lean on science when in the early days fertilizers in bags—or barrels in the earliest days—became available. In those days the soil was still black, and the easily handled fertilizers seemed to be just as effective as manure. Consequently many a farmer allowed his barn to become clogged with animal manure while he hauled in commercial fertilizers from the nearest railroad station. It was a matter of convenience, possibly; or a saving of time. Characteristically, American farmers have sought always the method that is easiest or most convenient, even though the well-being of their crop is not thereby best served. Land-area limitations elsewhere in the world make it impossible for farmers to relax their efforts to make the land produce up to its maximum possibilities. Here, there has been plenty of land. Thus, even at the cost of using more land, the American farmer could and did choose the easier ways. It is not altogether the fault of science that throughout the decades crop yields have decreased or barely maintained their former averages. It has been useless for scientists to try to persuade farmers that the more difficult way is better—as long as plenty of land existed over which they could spread their operations and manage to harvest enough total crops.

A few years ago, after a generation and more of scientific effort to help farmers improve their crop yields, surveys of the over-all results were made by agronomists in a number of the states. Their findings were not reassuring. It was discovered that, despite the actual adoption of many supposedly important improvements that should have effected increases in average yield per acre of 40 to 60 per cent, the actual average increase per acre was about 3 to 5 per cent. In other words, but for these adopted improvements, the crop yields would have been far lower than they had been fifty years earlier.

The average yields of crops at the time this investigation was made probably were at or near the low point of the country's agricultural history. Since then yields per acre have been improving. Hence, while it was feared in some quarters a decade or two ago that we were approaching an era of famine—despite our tremendous spread of cultivated acreage in this country—trends that have developed in the interval give assurance that that particular goblin is not going to get us after all. That is reassuring; but don't decide yet that we are just in the outskirts of Utopia.

What threatened to be famine, as viewed a decade or so ago, now looms as just the opposite. Instead of food shortages, we are more likely to have surpluses on a scale not previously imagined. Too much food, like too much of almost any other good thing, can be disastrous. And that is the possible threat we face.

A century ago we would have exported such surplus products to countries that needed them, but that is not easily done now. In those days our farmers did not have to put into their crops as much cash expense as now; the crops could be sold at prices low enough to make them attractive in world markets. Today, the American cost of production has mounted so appreciably that—except for tariff barriers—the farmers of other countries could easily undersell American farmers in our own domestic market. As long as this is true we cannot hope to have here a profitable agriculture in any real sense. An agriculture that looks to government aid for support is a crippled agriculture; and we need but to look about us and see how production is managed in other countries to know that such is the case. Yet, so far as I have observed, nothing except government aid is being suggested as the way out of this dilemma.

Since the farmers of other countries grow their crops less expensively without the aids that are urged upon American farmers, even though in most instances they do not have the assistance of machinery, there seems every reason to suppose that American farmers (supposedly just as intelligent) should be able to do the same. Indeed, with their advantage of machinery, American farmers ought to so cheapen their operations as to have the advantage in low unit costs. This is but the common sense point of view that every legislator must consider before he approves further assistance to our agriculture. He may quite properly suggest that it learn the ways of foreign farmers who have not so burdened themselves with costs, and whose production per acre exceeds our own. Legislators who thoughtlessly continue support as a matter of precedent may later discover that, unwittingly, they were subsidizing, not farmers, but the people who sold farmers the stuff that made their production so expensive.

Assuming there were real need for further government aid to farmers, there still is no assurance that our government, burdened with postwar debt, would be able to extend it. Federal revenues in the postwar period may not equal those of the war period, or even those of the prewar period. Moreover, your Uncle Samuel has been the United Nations' treasurer to a large extent throughout the struggle, and may continue to be such for years to come. You and I have pitched in and helped keep the treasury replenished in seven War Loan drives. In the postwar period, the single item of annual interest on the bonds we have bought will be more than the entire government expense of a decade ago. Besides paying this interest, Uncle Sam will have to find the funds for current expenses and bond redemption. Personally, I think he will have to practice unusual economies in order to meet necessary expenses. The chance that agriculture will be able to continue receiving help from the government seems to me slim indeed. While we are still the richest nation in the world, according to report, that has become an empty distinction; it is too much like class distinctions among paupers. Our country has indeed traveled the road from rags to riches—and back again.

Diagnosis of our country's situation suggests the need for reverting to a lower level of wages and prices generally; this to be both justified and accomplished by general cutting of the costs of production. The most difficult part of such a readjustment, quite naturally, is the start. Nobody wants to take the initial reduction in income. Agriculture, particularly, feels it has always been discriminated against in every previous recession. This time, if agriculture takes its cue from farmers of other countries who regularly grow crops at lower unit costs, the way is open for our basic industry to take important cuts in market prices without any loss whatsoever.

In later chapters tentative suggestions are made as to ways in which agriculture can plug some of the leaks in its system while it leads off in the general easing of prices and wages, along with reductions in all kinds of production costs.

3

PENURY IS SELDOM FATAL

ATTITUDES toward the possession and use of money differ widely in various parts of the world. Few peoples anywhere are as dependent upon money, on a nation-wide scale, as are we here in the United States. In the rural areas of many countries money is almost unknown, yet people manage nicely without it. We couldn't do that here; certainly not if the emergency requiring a moneyless way of life came upon us without warning. We take money for granted, use it almost automatically, and are literally unprepared to do otherwise. On that account, people elsewhere in the world seem to us strange in many ways. Their strangeness results chiefly from their more complete self-sufficiency. They do for themselves, in interesting—often surprising, not to say embarrassing— ways many services we pay for here at so much per item. We buy our clothes; they often make theirs. Most of us buy our food; they almost universally grow theirs. We send our clothing to the laundry; in some cases they wear theirs to the nearest stream, strip, and in the clear running water cleanse both clothes and bodies at the same time. Their customs are as anomalous to us as ours are to them.

They feel as superior to us as we feel to them. People are just that way.

To us, the status of people who have little or no money is abject poverty. People in our own country who have neither money nor the "know-how" of self-sufficiency that characterizes foreign peoples are indeed desperately poor. And, since such people share the common desire to be like their fellows, their inability to ape their more prosperous neighbors makes them miserable. Our slums, rural and urban, are populated with such people. Knowing that our own people are much depressed by their poverty, we assume, without realizing the fundamental differences, that moneyless people the world over are similarly dispirited. In this we are seriously mistaken.

This is a mistake we make easily because we see other people from a special point of view which is possible only to people who have forgotten long since how to be self-sufficient. When we appraise the simple folk of other countries as unhappy because they cannot enjoy privileges we accept as commonplace, we endow them, unconsciously, with the same point of view that we ourselves hold. It is true that these people know little about sewing machines, tractors, and automobiles; but they do know many things we neglected to learn from our ancestors. They can find trees full of honey in the forest; or track down and slay game without firearms; they can feed a large family, on nothing a year, throughout a long lifetime, and without the necessity for a physician. Despite appearances we interpret wrongly, these people have abilities and capacities far beyond our understanding.

Such people are not poor in any real sense, even though they know nothing of many conveniences which, in a country like ours, money will buy. On the other hand, if an American were left to his own devices in their country, with no filling station, store, or shop anywhere from which to obtain supplies, he would be utterly helpless. He would really be poor, even with a large bank account.

Poverty is not essentially a lack of money; nor is the mere lack of money poverty. Rather, poverty is the lack of ability, in any given set of circumstances, to get whatever is necessary for comfortable living. Here, where life depends almost literally on a bank roll, the lack of that roll of bills brings one to grips at once with realities of an entirely different sort, which proves that, after all, money is but a means to an end. The millions of other world citizens who have never known much about money would be little better off than they are if they received a weekly pay check. In fact, stories have come out of the South Pacific to the effect that some of the natives see no point to cashing the army checks they receive. These natives are penniless—poor from our point of view—and don't know it.

If we need further evidence that poverty—in the sense of lack of money only—is no great handicap to people who have learned to do without cash, we should find it in the persistence with which Japanese troops, completely cut off from their own forces and stranded on islands in the Pacific, kept up their guerrilla warfare as long as their ammunition lasted. These enemy troops, who had never known the advantages of depending upon a corner grocery, subsisted easily on what they could find in the wilderness;

and, when given time, supplemented this with vegetables they planted and cared for. People who have thus grown up with the necessity for self-sufficiency were not easy to conquer—a fact we discovered during the warfare in the South Pacific.

By contrast, people who have always marketed for their groceries and have been dependent upon tradesmen of one kind or another for virtually all their needs could be most awfully poor, and with great suddenness, if the source of their income should cease for any reason. They would be far worse off than the most penniless South Sea Islander. They would lack both the money with which they have been accustomed to supply their needs and the life-long habits which make the poor native so independent of money. The unsophisticated native would not spend money if he had it. The American, cut off from his pay check, would be helpless without it. Yet the essential needs of both are the same.

Until quite recently all this discussion would have been rather pointless. The economic problems created by our participation in two world conflicts within a single generation have conspired with domestic problems we have neglected to solve; and the combined result of these maladjustments makes it highly important that every person in the United States be put on notice immediately that he could quite suddenly find it necessary to become self-sufficient here in the United States itself.

All of which is but to suggest mildly that you would do yourself a great favor if you immediately began planning how to provide for your own and your family's food for the next year or two. If this sounds foolish in a country as prosperous as the United States is, you should make inquiry among people who are familiar with public affairs and long-range trends. Many analysts are committed to the thought that if one wishes to be absolutely sure of his food supply in the postwar world there is but one completely safe procedure: "Grow it yourself." I have heard such a suggestion made on the Chicago Round Table and on two or three other network programs. Consequently, I am sure that in making the same suggestion I am but expressing a sentiment that is shared by many thinking people of the country.

Uncertainty as to what will happen in the years of peace ahead is almost universal. It goes without saying, however, that the money crops upon which farmers depended during the war period cannot receive the same emphasis that they did before. Our plans must envisage a market which will be confined more strictly to domestic demands, though the imponderable element of world food needs will have considerable effect during the next two or three years—at least until world agriculture has re-established the position it occupied before the war. It is not yet possible, therefore, to prescribe a course of action for the commercial farmer, but it is safe to say that the average individual must provide a "cushion" against adversity by becoming a good deal more self-sufficient, in the matter of food requirements, than he has been in the past.

All the postwar plans—even those of industry—prescribe as essential a liberal income for farmers. This is necessary because farmers buy a big fraction of industry's

total production. Great reductions in farm income would
be disastrous for some industries which depend upon
farmers for most of their sales. Yet, as shown in the above
paragraph, it is impossible to plan in advance for a smooth
dovetailing of wartime into peacetime agriculture. This
is one spot where, if Robert Burns were living, he might
say again—with appropriate vacuum-tube amplification
—that familiar line: "The best laid schemes of mice and
men gang aft agley," yet no whisper of criticism of war
plans can be tolerated, simply on the grounds that they
do not fit neatly into the transition period. They were made
for war, were adapted to war, and should not be expected
to blend smoothly into the transition from war to peace.
It is too much to hope for that farmers will not experience
some serious reductions in income as an immediate effect
of planning for war conditions beyond the actual end of
hostilities.

Suppose the peace brings with it large reductions in
farm income; and suppose that, as a result, the industry
in which you work, or from which you receive income by
way of dividends, is forced to curtail its operations greatly
—or even close down. Are you prepared in that event to
continue to live on the scale to which you are accustomed?
Few would be, of course. The important question is
whether you would be in position to provide for yourself
at all.

There is not the slightest fear that farmers and garden-
ers will fail to produce enough food. While there has been
an acute shortage of some of the most important foodstuffs,
this condition is admittedly temporary. There is now no

apparent reason why there should not be enough food produced during the transition period, and on into normal times. Thus, the point of this suggestion is not that there is likelihood of food shortages, but that disturbance to the income of farmers may jeopardize the marketing of industrial products from which you derive your income. That event would quickly make itself felt by everybody connected with the business involved. Consequently, it is but the part of wisdom for everybody who is not already growing his own food to begin to do so, or to arrange definitely with some food grower to provide food throughout any crisis that may develop.

It is conceivable, too, that in the critical years ahead we may actually experience what has often threatened, a transportation strike of major proportions. That is something nobody wants to see, much less those who are not self-sufficient for food; but we can no longer say that it could not happen.

Anything that endangers your food supply is potentially fatal. If there were not possibilities that your food supply could be endangered by the events of the years immediately following the war, nationally-known men who debate these matters on the air would not be so unanimously agreed that, for absolute assurance of a food supply, you should grow it yourself. I know of no one who does not fully intend to be absolutely sure of his food; but unless someone points out specifically the circumstances which threaten a food crisis in this food-producing country, few of those most seriously threatened will be aroused to activity. Failure to act soon enough, in anticipation of such a crisis, could easily cause serious trouble. Some of our finest people would suddenly find themselves without food —and without the means for obtaining food. Such significant possibilities are the justification for this chapter, which in other circumstances would seem pointless. Poverty doesn't slay people; but the hunger that could result from failure to anticipate the possible disappearance of foods from the markets *might*.

History offers little precedent in this country upon which to base the suggestions of this chapter. Because of that fact, many a thoughtful man will be reluctant to show concern about a change in his habits. Such an attitude could be highly dangerous. I hope this chapter serves to frighten people enough to make them investigate immediately the factual background upon which this argument is based.

Nothing would please me more than to find, when history overtakes prognosis, that this warning had been unnecessary; that all our worry had been in vain. It would be pleasant to find that, after all, our planning had been so well conceived that we could change back to peacetime conditions without a ripple in our economic structure. No one—not even the most optimistic of planners—expects any such Utopian conditions to prevail. Hence, as a safety-first hedge against the possible failure of business confidence, transportation, or governmental effort to prevent such a crisis, immediate action is indicated as essential.

4

STORM CLOUDS AHEAD

T HE GREEKS AND ROMANS in their prosperous times could have said truthfully of their countries: "There will always be Athens," or "There will always be Rome." Both cities continue to exist. The hills, plains, and coastlines are unchanged. Some of the very buildings upon which Greece and Rome lavished their best talent still stand. Even the people who now inhabit the areas are descendants of those illustrious ancestors. Their form of government has changed greatly, but Greece and Italy still cradle such civilization as has survived.

The world's history is the story of peoples organized for mutual protection against their enemies. This is the essence of government. The form it happens to take is a detail. When governments disappear, they do so because they have reached crises for which they are unable to find solutions. The pure democracy of Athens reached such a crisis. The Roman Republic in its turn faced a time when such protection as the existing government could give was not enough. Their governments perished; but, obviously, not all of the people disappeared. It is not possible now to identify the causes that determined whether an individ-

35

ual survived or died in such a crisis; but it seems a reasonable assumption that those who did survive were able to control their own food supply.

Within the past few years both England and the United States have been face to face with circumstances which neither alone could have mastered. The most recent threat came from the Axis Nations, of course; but each nation has been threatened from within by forces which could eventually bring the downfall of the existing type of government and replace it by something similar to the types we have so recently fought.

Settlements with Berlin and Tokyo will still leave those internal unrests that have dogged our politicians for years. Indeed, the war's end ushers in the beginning of a train of economic unbalances such as we have never before experienced. Nobody knows or can know the answer for many of these impending problems. Economists to whom we usually look for predictions about future economic conditions are very quiet about the future we face. And no wonder. The variety and immensity of the grudges that are now to be reckoned with dwarf anything our country has ever known before, partly because the work incident to prosecuting the world's biggest war distended little grouches into big ones and fathered a lot of new ones we hadn't heard about before.

The farm problem is generally thought of as the one that will generate most trouble. A whole book could be written about its various phases. Industry offers an equal array of difficult problems to solve—not the least of which is the task of providing on a permanent basis some

55,000,000 jobs. Women workers who enjoyed cash incomes during the war continue to expect employment. Those who have returned from military service not only expect satisfactory employment but must be provided it. To keep men and women busy at satisfactory rates of pay while scaling down production from wartime levels—this is the reality with which America must cope during the coming ten years at least. It is not the short haul of immediate consumers' demand in peacetime that concerns us mainly, but the long haul of economic stability.

But the farm and industry phases of the general unrest are only the beginning of our troubles. Planners hope fervently that they are also the most important, for they ramify into every section of our population.

We must settle scores with many groups among the population—groups that have concluded that complete comfort and satisfaction is their right, regardless of circumstances. Every group that has become articulate through organization demands all the money it can spend. To tell these people that unlimited demands cannot be met is futile. To them such a suggestion would seem to be changing the rules during the progress of the game. They interpret their constitutional rights as guarantees of happiness—not merely as assurance of the right to pursue that fine state of mind. They feel cheated when told that there isn't enough money in the world to supply every person with everything he may want. This is no great exaggeration of the actual situation our government must face within the next few years.

Many students of government and economics feel that,

by weathering the crises that are upon us from agricultural and industrial reconversion, we can glide smoothly into the conditions that are desirable in the remaining half of the twentieth century. We hope that view is correct, for there are indeed enough threatening prospects in both agriculture and industry to frighten any but the stoutest hearted planners. Certain opposing trends must be harmonized both in industry and in agriculture if we are to avoid something approaching chaos after the backlog of civilian goods orders has been manufactured and distributed in the next few years.

Industry is one up on agriculture in the matter of this backlog of orders. There is no such accumulation of unfilled orders for farm products, except such as may be developed through chemurgic or other far-sighted planning against long-term needs. Chemurgists have been nipping at the heels of a tardily developing agriculture for a decade or more—with no noticeable effect. Now is the time of all times when agricultural authorities should welcome the help that chemurgy can give to a farming system that is rapidly becoming top-heavy through excessive production along strictly conventional lines. Farmers scarcely know how to turn away from the habitual food, fiber, and feed pattern that agriculture has followed since the very beginning. The time has now come for such turning; and the solution of some of agriculture's gravest problems may lie in just that phase of the possible activities. More about chemurgy in a later chapter.

Industry can be more deliberate in its adjustment to peacetime matters; for it will be occupied for a considerable period in satisfying the demand for civilian goods which could not be met during the war. Its workers during this period will receive their regular pay checks, and there should be a minimum of difficulty.

Agriculture faces an entirely different situation. Customers for food cannot file their orders and await delivery of the manufactured delicacies months hence. There is not, therefore, a backlog of unfilled orders awaiting the attention of farmers. Moreover, the crop trends initiated during the war must be readjusted somehow, but nobody seems to know just how, to the greatly diminished peacetime consumer demand. Regardless of what direction agricultural planning takes from now on, the disposition of the final abortive wartime crops presents a problem that must be solved with a minimum of dissatisfaction all round.

Satisfactory disposal of these final war stocks of food and fiber, along with the last war-planted crops from our farms, will require a degree of finesse beyond that required for any previous transactions. If you wonder what will be different about this event, consider the fact that both agricultural and industrial leaders, in their postwar planning, have established the questionable "principle" that in the postwar period there must be no decrease in incomes for farmers. On that account, the peacetime sale of agricultural products will be watched closely by everybody concerned. And, since the supply of farm products is now expected to be far in excess of any normal peacetime demand, consummate skill will be required to dispose of the entire quantity without depressing the price and the farmer's income.

Doubtless, everything will go off all right. It will be done under government supervision and control, of course; and there is ample precedent for subsidies, support prices, or outright government purchase of a portion of the offerings, if artificial stimulus is required to avoid depressing prices. After this final clean-up of wartime stocks, what then?

The most pressing problem that confronts agriculture after the war trends of production have been changed will be to devise ways to avoid the annual recurrence of mountainous surpluses. The stage is all set for the repetition of prewar production of surpluses—especially of cotton, corn, and wheat—on a bigger and grander scale. Several factors tending in that direction may be cited:

(1) Some 800,000 farm boys were in the various armed services. Most of those discharged are starting in where they left off when their war service began. Nobody has wanted to urge them to do otherwise, of course.

(2) Other multitudes of young men who have been in the services and who know nothing about farming have come to the conclusion that rural living is the way to the comparative peace and quiet they have dreamed of enjoying after war service. Every official and unofficial source of information about farming is now being bombarded with requests from these men. Some are just starry-eyed dreamers, of course; but not all of them are such. Many have well-laid plans for tying in an industrial job with life on a few acres where they can have a cow and chickens. They present their plans with requests for criti-

cism or suggestions. In many cases these plans reveal a surprising acquaintance with the pitfalls involved; the young planners have apparently done a good deal of careful reading and perhaps have had consultation with friends who know farming.

(3) Uncounted thousands of city people were poised for a move to the country when the war began and halted their plans. The inconveniences of rationing, increasing dissatisfaction with city life, perhaps, too, a few years of successful experience with home-grown food from small gardens—these influences have further confirmed the desire of many people to move out where they can grow part or all of their food. In the new era of peace these people will find places for themselves in the areas just outside the city limits, where they can enjoy the pleasures of country living while still receiving a regular pay check from a city job. The mail bags bulge with inquiries from these people. They want to be sure their plan is safe.

(4) Others in great numbers—war workers who moved into the city from farm or village to earn war wages while the earning was good—are investing some of their savings in small landholdings where they can settle. These people will produce their own food, if not a surplus for sale.

All of the above definable classes of prospective new producers of food loom as a threat to an agriculture already potentially overmanned. In the year 1944, with 15 per cent fewer people working on farms, and with an acute shortage of the machinery considered essential to normal

crop production, the total national production of our farms
was 36 per cent higher than that for 1939. Indeed, each
year of the war witnessed production that was record
breaking by comparison with the next preceding year.
Since even in 1939 we had corn in many parts of the country that we didn't know how to dispose of, it should be
clear that the business of farming cannot absorb new thousands of people who will at least shrink the potential
market by the amount of food they produce for themselves
—if they do not go further and grow additional food for
sale.

Note, in the above, that farm production *for the abnormal needs of war* has been accomplished while farmers
were shorthanded and poorly equipped with machinery.
The miraculous character of this situation must be accounted for; because, even without the potential threat of
new people in farming, we already have a situation within the industry which can wreck the farm price structure
as soon as the war demand has eased. How can this mys-
teriously abnormal production under handicaps be accounted for?

I think there are some factors in this phase of our
farming that have not generally been given the credit they
deserve. Many who discuss the phenomenon believe that
embattled farmers were aided by weather conditions con-
tinuously more than normally favorable throughout the
country. Possibly this is the explanation we should accept. I know of sections of the country, though, in which
there has not been a season within the past five years when
farmers did not experience what they considered serious

droughts, and droughts usually result in lower, not higher yields. Infrequently, it is true, yields are aided by droughts; particularly during the early part of a season when crops should be extending their root systems, and again at, or preceding, harvest time, when rain could prevent proper maturity or cause spoilage. On the whole, though, it seems to me that we may account for the increase in total farm production in ways more plausible. Weather conditions favorable in one area are likely to be cancelled out by equally unfavorable conditions elsewhere. We have other potential factors that I have never seen mentioned as possible causes, the effects of which are uniformly toward production increases.

The most important of these, perhaps, is the cumulative effect of yield-per-acre increases on farms where "soil conservation practices" are being applied. The Soil Conservation Service has been in operation for more than a decade. Though it began in a small way, it has added territory to its sphere of influence each season. It has modestly claimed that farmers who follow the recommended practices achieve yield increases averaging 25 to 30 per cent the very first year the new methods are in effect. Since these effects are cumulative, and since the amount of acreage involved each succeeding year has also been increased (and will continue to be increased with each future year), it seems entirely fair to suspect that the total farm production of the past five to ten years has been importantly increased by the operation of the Soil Conservation Service.

Records of the Service easily could fail to include a

great deal of increase that properly may be accounted for by the indirect influence of the example set by farms where these practices are followed in co-operation with the Soil Conservation Service. Many farmers hesitate a long time before they sign an agreement to co-operate. Some never sign. Yet no farmer who has sat on the line fence chewing a straw while he watched his neighbor harvesting a superior crop by reason of such co-operation can resist the temptation to follow the better practices—even though he may never become a bona fide co-operator by signing the required agreement. Soil Conservation Service records could not possibly include all the increased crop yields that result from the oblique co-operation of the straw-chewing gentry. And it would not be strange if investigation would disclose that this "sphere of influence" included more total production increase than can be accounted for among the regular co-operators. Human nature and stubborn politics, in other words, may be credited with part of the farmer's threat of extinction by surpluses.

Soil Conservation Service workers believe that quite a fraction of the production increase must be accounted for by the planting of cultivated crops on land which has been in grass. Perhaps there has been considerable such acreage. There has also been a great deal of formerly productive land that has *not* been used for crop production since the operator was called to the colors. These abandoned acres may equal in area the grassland that has been cultivated. I am not aware that there has been a survey to determine the balance between these opposed factors. It may be said, though, that much of the abandoned land

lies in the Northeast and the Middle West where, normally, the marketable surplus of crops is produced. The removal of this land from production, then, constitutes an
important reduction of the potential output of American
agriculture in recent years.

These, then, are some of the clouds that are boiling up
to disturb the dreams of farmers and confuse those who
are planning the future of agriculture. As remedies for
this situation, some strange suggestions are being made.
One economist, at least, has come up with the bright idea
that three million people must be removed from agriculture. This man participated recently in a nation-wide
broadcast devoted to the subject of planning for the postwar period. The double-barreled export suggestion—both
industry and agriculture exporting to customers presumed
to be able to pay out money they don't have—is made
more often than any other. Sixty million jobs . . . guarantees by the government of *full employment* . . . continua-
tion of government subsidies and controls, for agriculture
particularly. All these are among the dreamed up "plans"
being suggested. One lone man, as far as I have been able
to learn, suggests a possibility which seems to me to offer
real hope: Ladd Haystead, the farm editor of *Fortune*
Magazine, believes it possible that farmers may cut their
costs of production enough to enable them to market at
least *more* of their crops than would otherwise be possible.
I agree with this thought and will suggest in later chapters
ways in which farmers may accomplish decreases in *production*, as well as in production costs.

I believe the reader will agree with me that, if as many

factors as I have suggested threaten to increase our present mountainous surpluses, we cannot afford to depend upon the thin reed of government controls—which is the "first aid" most writers and speakers think of easiest. Beneficial as government controls have been in the past, we know that they have been far from perfect; moreover, they have been directed to surpluses in miniature, as compared with the possible overproduction now envisaged. Even though we may need to continue, and perhaps strengthen wisely, some of the governmental controls, I believe we will err seriously if we do not also plan to make major changes in our general crop-production scheme with a view to reductions both in the volume and in the unit-cost of our production.

5

WHEN THE STORM BREAKS

IF HISTORY offers a trustworthy precedent from which to blueprint the new world that is to follow upon the war just closed, I am not aware of it. It is certain that nothing in history closely parallels the world-wide situation that prevails today; and in this situation the United States is in very special economic contrast with most of the other nations. At no time in history has so great a number of people in any large country been fed by the labor of so few. And no other large country at the present time supports so many people by the efforts of so few growers of food. Our problems with reference to the future of our own country, therefore, are entirely different from those of our world neighbors; and, in our efforts to co-operate with other nations for the maintenance and promotion of peace for all future time, we must manage to resolve our own serious problems in such a way that we can do our part in a world organization.

In some ways the economic relations between nations are as simple as are those between individuals. It would be easy to overwork such a comparison, of course, but it can do no harm to consider some of our national and international difficulties as they may be illustrated by what we know about individual relationships.

The United States is the richest nation in the world— even more so now than it was before the late war. Not only is that true, but we also are creditors of many of our allies for great sums of lend-lease aid, feeding of suffering populations of the reconquered territories, etc. But we cannot really consider ourselves very rich nationally any more, for most of what we have poured into our part of the war effort, as well as what we have lent or furnished to our allies, is being charged on the books against the future of ourselves, our children, and our children's children. Our indebtedness to every war-bond holder in our own country is now so great that our opulence is chiefly a matter of bookkeeping. Yet—and this is extremely important, it seems to me—we retain the same wealth-consciousness that we had before we became entangled in World War II.

Our interrelations with other countries must be influenced by this mental attitude. The wealthiest nation of the world cannot be expected to negotiate with its less fortunate contemporaries in the same spirit that actuates them. Certainly individuals could not do so; and it would be rash to expect that the representatives of nations could completely avoid being influenced by the environment in which they have lived their entire lives.

For illustration of our *actual* relation to others financially, the following may serve: A group of small boys playing in the street suddenly find themselves hot and thirsty. With the exception of one boy, they have but a few pennies among them. This one boy has a couple of dollars, part of which belongs to his sister, but that is a matter between himself and her. In their thirst the boys file into the corner drugstore and fill up on sodas, paid for chiefly out of the fortunate boy's funds. When the time comes for settlement, the "financier" of the crowd has three cents left, while the others have nothing and owe him for most of their sodas. He is richer than they, but his actual wealth is a far different kind from that which existed when the festivities began. How he will settle with his sister, of course, is not their affair.

That, roughly, depicts the relative financial situations of the United States and its allies at the moment. The fact that our effective "mortgage" against the total appraised value of the country's wealth may be dangerously high doesn't improve the picture. A small boy in the situation described would feel very poor. Not so the people of the United States. We still cherish the idea that we are the richest nation in the world and intend, come what may, to continue living in the manner to which we have become accustomed—or better.

A fair approximation of our mental attitude towards the rest of the world and our co-operation with it economically is more like this, if I do not misjudge the viewpoint: In a crowded hotel it becomes necessary for an American of great wealth and position to share his room with the King of Hobos. Our socialite is socially conscious and is willing to split the expense of the high-priced room equally, or to make any arrangement agreeable to his roommate. The other, being also an American with a streak of independence in his mental make-up, suggests that they share, instead, a fifty-cent bed he has located in a familiar flop-house. You may complete the episode, following the practical lines of development that are inherent in the situation. I can't visualize the transfer to the flop-house. Perhaps you can.

As matters now stand for our country financially, we still have the wealth complex without the wealth—not yet having realized that we are no longer wealthy in any real sense. Crude and elementary as these illustrations necessarily are, they should make it possible for us to realize that we are seriously handicapped in our attempts to co-operate with the other nations of the world on an equal basis. There are many evidences that we still retain our habitual confidence in money as the one and only cure for national ills—our own and those of our world neighbors. And it should be obvious that, until we think more realistically about finances, we shall not be able to work with other nations on an equal basis. It is highly important, then, that we immediately come to our senses financially, in the interest of the success of the present attempted peace organization.

The most serious hurdle in the way of scaling down our money complex is the stubborn fact that our wage and price scales are, and for decades have been, far higher than those of any other country, with the possible exception of Canada. And this difference has been made much more complex by the introduction of governmental controls during the war period. No way exists for judging the extent to which both wages and prices would have risen if nothing had been done to prevent such increases.

Yet, however painful the process may be, if we are to associate as equals with other peace-loving nations with far lower scales of value, *we must begin somewhere,* and the way is through drastic reductions in our own scales.

The most difficult part of the whole operation is making the start. Nobody is willing to accept less for himself, if everybody else is to receive the same as before. Our problem for initiating the scaling down is to find the place where a beginning can be made without precipitating disaster. This is precisely the problem for which this chapter tries to suggest the solution. And, at the risk of being indicted for treason against the sacred doctrine of high price and wage scales, I ask careful consideration of a procedure which need not involve disaster, and which could be followed, it would seem, by concessions in other quarters —concessions that would be made less painful by the initiation of this first reduction.

Preposterous as the suggestion will seem to farmers and to other students of agriculture, great concessions can be made by agriculture if farmers will adopt methods equivalent in principle to those followed in other countries where yields per acre are higher than in the United States. Many readers will doubtless be surprised to be told that American farmers produce less per acre, on average, than the farmers of almost any other country. We can account for such differences in one of three ways, it seems to me: either

(1) the soils of this country are inferior to those of other countries where larger yields per acre are obtained; **or**

(2) American farmers are less skilled than the farmers of these other countries; or

(3) the methods followed by American farmers are not based upon principles as sound as those followed by farmers elsewhere.

I can think of no other possible explanations. (It should be added, though, in defense of the American system that our farmers do produce far more *per man* than do farmers anywhere else in the world.) Now let us examine these possible explanations separately and see whether we can arrive at a tenable theory to account for the apparent inferiority of the American system on the score of productiveness per acre.

We know that our soils did produce several times as much per acre when they were first cleared as they do now. Indeed, in earlier days we fed from these soils many millions of people living abroad. Our machinery made it possible to produce food from the fresh, lively earth so much cheaper than it could be grown in other countries without machinery that we monopolized world trade in food for many years. We must surely discard the assumption that American soils are basically inferior to those of other countries.

No one would argue against the skill with which the American farmer practices his chosen methods. It is too obvious that he accomplishes what he sets out to do. He gets done the most it is possible for a man and his machines to do—within his chosen range of methods.

But, what of his methods? Obviously, those that relate to the handling of the soil are the ones that need to be considered; and it must be said of them that, by comparison with the methods used in other countries, they cater to the convenience of the farmer rather than to the well-being of the crop. In other words, the American farmer uses his machinery for covering territory, even at the expense of yield per acre. He plows deeper, the more completely to rid the surface of his soil of what would interfere with his later operations. Most of the later operations are performed with sliding, instead of rolling, movements; thus it is essential that the exposed soil surface shall present no obstructions that could cause loss of time in tillage. In behalf of speed the American farmer sacrifices his most valuable aid—decayable organic matter, which, notably, is hoarded by peasant farmers everywhere.

It turns out, then, that the one point upon which the American farmer has considered himself superior to farmers of the rest of the world is also the one which accounts for the gradual decadence of his soil. He wastes with his plow what he must atone for with his purse. And that expensive waste, all of which could be saved by a suitable change of methods, accounts for more than half of the per-unit production cost of every American farm product.

What I am suggesting, in short, is an immediate and drastic change in both thought and practice regarding the soil. Such a change will return to our farmers their former tactical advantage in world markets—a point that now worries our postwar planners. Changes in method which

are to be suggested in later chapters will make it possible
for American farmers to produce as much per acre as the
Chinese do, without the necessity for befouling the atmos-
phere while doing it. Moreover, these results can be
achieved without the temporary but appreciable losses
that must usually be anticipated in other forms of "reconversion." We will simply produce as much—or more
—with smaller outlay of money, less labor in some instances, and in some cases with less machinery. Every
item in the suggested program has the stamp of approval
of science, or has had it within the past twelve months,
is entirely workable, and can be started on American
farms at the appropriate season within the next twelve
months.

Obviously, if American farmers in practicing such a
regime spend less cash, somebody who has previously
been supplying the service that is to be discontinued will
receive less income. It could not be otherwise. But, if we
are to achieve our announced intention of reducing the
cost per unit of American farm products as much as possible, the farmer's cash outlay must be reduced at every
possible point. There is no way to avoid cutting across
commercial interests in such a move; but that, too, is not
new to America. We did not introduce the automobile,
nor any other of the thousands of new conveniences this
country has enjoyed, without cutting somebody out of
income he formerly had. The stock suggestion to every
person who conceivably could lose income by reason of
this suggested improvement in our farming is that he
should make absolutely sure of his food by growing it

himself. And that suggestion, by the way, is expressed and approved by most analysts when discussing our future. This, in itself, is an indication of how fearful our country's leaders are that we shall not be able to avoid trouble in the transition from conditions that have recently prevailed to those that must and will prevail in the next ten years.

This change in agriculture, it seems to me, should be started as soon as possible. It cannot be made quickly, for even those who are to educate the farmer to its necessity and advantages must themselves first be shown how essential it is that we lower our basic levels of value. On this account, no time should be lost deciding upon the necessary policy changes and organizing the educational campaign for implementing them. If we should succeed in launching agriculture upon a less expensive, more productive plan within a year's time, we should be able to increase the ease with which we move from war to peace in industry and business. With agriculture already on a low-cost basis, it would be far easier for everybody else to scale down his economic requirements.

Such procedure, by lowering living costs for everybody, will pave the way for progressive lowering of wages, costs of industrial products, and everything else. It is easy enough to see that enough of such lowering of costs would cure a lot of headaches connected with exporting our surpluses, both agricultural and industrial. And, if we can go into the export market on an even basis with the rest of the world, our postwar commercial problems may very well cease to exist.

Nothing can ease the tensions of our lives so much as a reduction in the cost of food. With assurance of abundance of cheap food, industrial workers will sleep better; they will need correspondingly less rigid assurances against an idle day occasionally; employers, thus freed from incessant and increasing demands for higher wages, will be keen to improve their competitive position in the market by reducing the prices of their products. Then we can all buy more of these products, thus increasing the hours of employment for labor, which, in turn, will increase labor's ability to buy still more of the products of industry.

There are, without doubt, many problems to be solved in the realization of this basic change in our economic system. None of these difficulties can be important enough that we can afford to let it stand in the way of our initiating this change. The way to world peace will certainly be improved by any move that brings our national economy nearer to the general level of the other nations of the world. Of that I think there can be no doubt.

6

THE TRUE ECONOMIC BASE

W E HAVE HAD a childlike faith in money as the universal solvent for our economic distresses. Our first thought in a national crisis has always been to provide for the *expense* of its solution. So accustomed have we become to "oiling" economically troubled waters by the money treatment that we have completely lost the tradition of simple, inexpensive cures. Spending is our way of keeping everybody supplied with the cash we consider necessary for future prosperity. Yet other nations, because they have neither our vast pool of wealth nor our traditional faith in money, must find cheaper solutions for their problems. We have not yet studied carefully these canny financial habits of our world neighbors, but we shall soon have to adopt them.

Our failure to sense the error in our dependence upon money as a cure-all for every national economic ill may be accounted for in part by another fact. In addition to always having money, we have also had abundance of food, save during World War II when it was necessary that our production be shared with our allies and with the populations of devastated countries. Never have we had to face here the basic alternatives of food *or* money. Here it has always been food *and* money. The difference is significant. Some of our citizens now know how unimportant money is, relatively. They have been in Axis prison camps. No alternative was theirs. They had little of either food or money. The one was vital, the other didn't matter, there having been no chance to purchase food. Their experiences ought to help us sense the need for realistic thinking in approaching our future economic problems.

Fortunately, too, we have for study the success of our neighbors in world trade, where we in late years have been at a disadvantage. Despite our supposed advantages of mass production in industry and machine methods in agriculture, we have gradually become unable to compete on an equal basis and have had to resort to numerous artificial devices in order to avoid being completely shut out of the trading in some commodities. It is not pleasant for a nation with proud traditions of past success in trade to be faced now with prospects of ever increasing and unmarketable surpluses in both industry and agriculture. How could it be that our neighbors, mostly underprivileged by comparison, could compete successfully with us in fields in which earlier we had been the leaders of the world?

Our world neighbors, we can now see, have had advantages of another sort. In fact, the advantages they possess are such that we have thought of them as disadvantages. They have had far less industry, far more agriculture. Instead of three-fourths of their people being supported by the food production of one-fourth, it has been

the other way around. In the past they were "hick" nations by comparison; but they could outtrade us. Food,
expensive here, was cheap with them. In many of our
neighbor countries there was not even a commercial agriculture in the sense in which it exists here. Most industrial
employees grew their own food; if the factory was shut
down, the fact made little difference to them. Such conditions lacked the excitement supplied by our country,
but were important aids to stability. Nobody was forced
into our kind of pell-mell existence. There was leisure.
Farmsteads could be built of stone for permanence. These
conditions, known to us only by report in the past, have
recently been observed by millions of Americans abroad
—eye openers to men who like to quip that in America the
only permanent thing is change!

We have always interpreted our continual succession
of new styles, up-to-date models, and extra features as
improvements. Sometimes, indeed, they are, but not always, and not necessarily. Certainly nobody can think
that change for fashion's sake is an improvement in the
usual sense of that word, yet a large part of the merchan-
dising of the United States in its heyday came from pur-
chases we made because we had to keep pace with the
Joneses. In this respect we differ from most of our world
neighbors. In the old days Paris profited from selling a
seasonal series of women's new fashions. Yet the average
Frenchwoman had neither the means nor the inclination
to indulge such a fancy. Why? American women had
money and were attracted by every new style they *thought*
Frenchwomen were wearing, whereas Frenchwomen—

the majority of them, certainly—wore what they had quite contentedly, helped grow their own food as a matter of lifelong custom, and salted away the profits on the new feminine finery. To each, the women of the other country were a little bit queer.

All the differences in customs of foreign countries may seem unimportant; but, in reality, many, if not most, of them are basic and revealing. We can't understand the British, who must have their spot of tea whether business goes on or not. That time-honored pause is tacit evidence that life among English people is less tense. Likewise, the Mexican siesta. As a people we are inclined to feel superior to such people because we "save" time and they disregard it so shamelessly. We neglect to wonder why time is less important to them. There is a very good reason. Thousands of British families, and most Mexican ones, have little gardens that feed them. Food-security is in the kitchen garden. And food-security is basic. The assurance that day by day, come what may, there will be food—that assurance, which may properly be called food-security, is the practical equivalent of two of our sought-after freedoms: freedom from fear and freedom from want. How unexpected that we should find that these have long been the possessions of many foreign peoples while we here are still searching for the formula for acquiring them.

Food-security—freedom from fear and want—these go with tranquillity. Food insecurity, even when not actual but only sensed as a consequence of remoteness from the land, breeds a subconscious tension. Some such influence must be responsible for the tensions that go with our fast-moving civilization.

Though food-security is basic and compensates often for the lack or uncertainty of money income, it would be easy to draw too sweeping conclusions from the discovery that our population is among those having least assurance of daily bread while most of the supposedly backward countries of the world are food-secure. We on our part are well organized to compensate for that lack of natural food-security by a system of commercial production and transportation of food that is unequaled elsewhere in the world. Nowhere else do solid trains of food travel at express speed three thousand miles to supply the tables of hungry people. Commercialization of some phases of our agriculture is well nigh as complete as is that of our transportation and industry. Only this strictly American development can account for the fact that one-tenth of our people grow the food that they and the other nine-tenths eat.

As long as nothing disturbs the delicate balance essential to the smooth operation of such an intricate system, we are just as food-secure, in effect, as we would be if each of us grew his own food. Throughout the long period of our country's economic development, the evolutionary process has been so gradual as to be unnoticed except by students of such matters. With plenty of money always to be had for progressive ventures, it was easy to establish the commercial phases of our agriculture; just as it was easy to build railroads, steel mills, huge grain elevators, skyscrapers, and the other necessary "tools" of our highly

developed economy. As our agriculture developed into
its bigness, along with similar giant developments in industry, business, transportation, and communications, few
could foresee that sometime we might find that our whole
people had lost something for each gain made by agriculture. Now we are approaching the time when we must
take stock of our unparalleled situation. We need to make
sure that we can pass through whatever crises lie ahead
without a fatal disturbance to our inverted economic pyra-
mid. Everywhere else in the world the nutrition of populations is broad-based. Many people feed few non-pro-
ducers of food. Here the few feed the many. And this
situation is obviously unstable by nature. Our problem is
to maintain the necessary balance; and our most serious
economic question just now concerns the possibility of
achieving this without serious maladjustments.

This is by no means the first crisis we have faced. Indeed, in at least one respect it is no crisis. We have absolute assurance that there will be plenty of food. We do not
have corresponding assurance that nothing will happen to
make the distribution of that food difficult or impossible.
And the distribution problem is not one of transportation.
It is strictly one of economics. We have learned already
that, even under government controls, food moves most
freely to the market that yields the most profit (or the
least loss) to the producer or handler who does the ship-
ping. That is a situation we are certain to find repeated
frequently in the peace years, whether food moves under
government controls or not. In fact, such behavior shows
that the human bargaining instinct never misses an oppor-

tunity to get the most profit possible. The law of supply and demand continues to exert a tremendous influence regardless of how strenuously government tries to neutralize it. When driven to cover, it reappears in what we call black markets.

This is by no means intended as criticism of any sincere effort to prevent food costs from rising. The public is entitled to cheap food always—the cheaper the better—but efforts to prevent rising prices by *ceilings* is a good deal like clapping a board over the chimney top to keep the smoke in. It is best not to start the fire in the first place. The force that sends food prices up is the cost factor in every phase of the production and handling of food. Why not tackle that side of the question? Real results will crown such effort.

No, government can't control the human nature which makes every producer or handler of food sell in the market that yields the biggest net profit. Neither can it repeal the law of supply and demand which determines for every locality the normal range of prices. It could quite effectively educate inefficient producers and handlers of food to less costly methods, with the result that they could then reduce prices without loss of the profits to which every producer and tradesman is entitled.

The educational effort designed to show that it is both necessary and possible to reduce production costs should include lessons, too, for that school of postwar planners who still hold to the theory that, come what may, wages and prices must be maintained on a high level in order to avoid destroying our high standard of living. This wide-

ly held philosophy has always worked, and will continue
to work for everybody *except* the ultimate consumer. That
personage, be he the purchaser of a tooth brush or an automobile, can purchase only as many items as his income
will permit. The higher our scale of prices the less he can
buy of food or any other necessity. Moreover, if he can
grow his own food, so that he need not spend money for
it, he will be food-secure and will have just that much
more to spend for the multitude of other things he may
want. Or, even if he cannot grow his own food—as many
in this country necessarily cannot—the lower food costs
will serve to a degree the same purpose of leaving him
more income to spend for other goods. This apparently
is a point of view most of our economic planners have not
even thought of. So insistent has been the high wage and
price propaganda of those who stand to profit by high
levels of value that few have dared question its validity.
When once we realize that the ultimate consumer group
includes every man, woman, and child among our population, we should be able to see that the total volume of
business would be upped by anything which reduces re-
tail prices.

   Traitorous as it may sound to commercial food growers and to those in the food trade, the more completely our
entire population can become food-sufficient by growing
gardens, the more cash can be available to spend for things
other than food. Also, food-sufficiency is obviously our
best refuge from violent ups and downs of wages and
prices. We tremble in our shoes now when our income
is threatened *only* because it is food, the only means of

life, that really is threatened. The direct threat is to money; but the indirect threat is to the food that money will buy. Self-defensive production of food is the effective remedy for the psychological upsets that follow great reductions in wages or great increases in prices.

So strange is this suggestion by comparison with our customary philosophy of American life that we shall not easily conceive the changes it will work in our scheme of things. We shall live more like the people of other countries, in that we shall gradually scatter our city populations so that the adjacent rural and suburban areas will be more densely populated by the overflow. Increased family production of food among people who formerly bought all their food will reduce the volume of in-season vegetables and fruits carried into the cities from commercial producing centers. Commercial food producers will find it necessary to diversify their operations to include the production of chemurgic crops, drug plants, or fancy food crops too difficult for the home gardener. The extent to which self-defensive food production may transform our whole agricultural and industrial scheme is impossible to estimate in advance; but it is safe to predict that a generation from now we shall wonder why we waited to be driven into a kind of life so much finer than anything we had ever known within the memory of anyone living.

The transformation of our economy from what we think of (incorrectly) as money-based to one that is more strictly based upon food-sufficiency ought to bring us important advantages without the necessity of sacrificing the best of our present system. While the food-based scheme of life of our world neighbors in many areas includes extreme poverty, it does not follow that we need to sacrifice the economic advantages we have enjoyed. The poverty-stricken peoples of the world are not so because they are food-sufficient. They are poverty stricken because the preceding generations that occupied the areas did not build up cash reserves like those that came to us almost automatically as the result of liquidating our assets of natural resources. We should be able to revert to food-sufficiency without serious loss of the advantages this wealth made possible, providing we do not delay our move in that direction until we have completely used up our heritage of accumulated wealth.

7

NEW PSYCHOLOGY, NEW METHODS

THE MOST DIFFICULT of all habits to break is a habit of thought, and the most painful of ideas, a new one. Hitherto, our thinking has been patterned upon a low national debt structure—in fact the Federal income tax has become a fixed policy only within the past quarter of a century, and, even so, there have been those who thought it could be abolished after the debt from World War I had been paid off. Today, we can forecast a national debt of some three hundred billions, or just over $2,200 for every man, woman, and child in the United States. We cannot continue to assume, therefore, that we now take up where we left off when the late war began.

Our world is changed in ways that have little to do with the new wonders of atomic energy or the near perfection of jet propulsion. The harsh fact is that we must now adjust all of our thinking to the reality of a debtor's position. It was one thing to be in debt nationally to the rest of the world, as we were before World War I; it is quite another to dismiss the burden which is now upon us, despite the bland pronouncements of the school of free-spending economists who conceive an internal debt as of no matter. On the basis of an anticipated debt of three hundred billions, the interest charges alone will be more than twice as much as President Coolidge's largest annual budget in a period of unequaled prosperity. And let there be no mistake about it: we shall have to pay the interest at least or face the consequences.

If we add to the amount of interest the necessary sum for current annual running expenses of government, the annual budget will run to eight figures. This, it must be remembered, is not the debt, but a rough estimate of the cost of writing it off. And there are other costs which have never been negligible—state and local taxes, principally, though extraordinary costs like those for unemployment relief, old age assistance, public works, and a number of others cannot be forgotten.

It follows from all of this that our psychology needs changing. That it will be difficult, goes without saying. In little more than a century and a half of our national existence we have telescoped the whole of the long and painful evolution which has been the lot of other countries whose roots go back a thousand or more years. Swiftly, dramatically (and perhaps tragically), we have arrived at the position that many a country before us has occupied. Opportunity remains, and natural resources, and the normal drives and incentives, but we are, for the first time in our national history, poor. Hard work, tight money, and reconciliation to reality are the conditions ahead of us.

That these factors are even yet not apparent at the higher levels of government becomes clear when we consider that government departments are in some instances thinking of expanding programs, rather than contracting them. Indeed, there is the prospect that departments will increase personnel in the next two to five years. It seems to be forgotten that, whatever the demand of consumers may be (and all forecasts are high), lush times are really not for us—not as long as we are overdrawn by three hundred billions.

If our thinking can be changed sufficiently so that our true position is apparent to us, perhaps our methods may be changed too. And without a change of methods, we cannot hope to meet our responsibility. We may become more critical of traditional ways of doing things, and even of highly propagandized but ineffectual plans and panaceas. Some of the most inappropriate and captious of current "reconversion" ideas have to do with agriculture. Business in peacetime may be "business as usual," but not so agriculture. The ancient misdeeds of agriculture have played too important a part in creating the conditions for future instability to be passed over lightly. Yet evidence accumulates that even the leaders of agricultural "reform" do not recognize the handwriting on the wall. Let us take an easy example:

In the May, 1945, issue of the *Country Gentleman*, Dr. Hugh H. Bennett, Chief of the Soil Conservation Service, outlines the supposed need for banishing straight-line culture from all of our sloping land. Much has been said and written in recent years on this subject. The straight row has become anathema, virtually by government decree. Writers have lauded the artistic beauty of the contoured farm in illustrated feature articles in the

press. So effective has been this indoctrination of the
public that the casual reader has come to accept as necessary this proposed face-lifting of our farms. His thinking
has been done for him at so much per word. He is no longer
conscious that water could not run off a leaky roof. He has
ceased to think such things through for himself, since we
have scientists to do the thinking for us. Consequently,
most of the public is all for whatever is recommended
to keep the soil in place.

The truth is, however, that gravity still pulls rainfall
into every square inch of soil surface that is not too dense
to admit the water. (Note that in all the writing about the
necessity for engineering of various kinds to handle the
rainfall from farm to stream, nothing is said about the
possibility of making the surface of the soil porous.)
Water can no more run off a highly absorbent surface
than it could run off a sponge—until the internal space
had been filled. This, of course, is logical; and farmers
everywhere who have tried mixing plenty of organic matter into the surface of their soil have found to their delight
that one of the unexpected effects is that the rainfall no
longer runs off. Many farmers write me about this after
having read my *Plowman's Folly*,[1] in which this idea was
set forth and urged upon farmers.

It should be obvious that, since almost every farmer
has implements he can use to so treat the surface of his
soil, all he needs is to be told how to create this spongy
surface on his cropland. Once this has been done, even
on a few farms, a start will have been made toward intro-

[1] Norman, University of Oklahoma Press, 1943.

ducing this less expensive and simpler method of solving the country's erosion problem. The Soil Conservation Service already has demonstrated that, due solely to the extra water that soaks into the soil, contoured fields produce bigger crops. That is exactly what would be expected to happen, of course, because all plant food must be in solution to be of any use; within ordinary limits, the more water the soil takes in the bigger crops it can grow. And it turns out that when plenty of absorbent organic matter has been mixed into the surface of the soil the farmer gets, not only the increase due to greater water intake, but several other advantages.

Unless too much rain falls all at once, soil that has been filled with a great quantity of organic matter takes in all the rain, instead of allowing some of it to run off around the contour. This results in *an even bigger crop increase* than is accomplished by the contouring itself—because there is greater water intake. Besides, much of this water is held in the organic matter, assists with its decay, and takes with it a heavier load of plant food when it is taken up by plant roots. These multiplied advantages result sometimes, especially on poor land, in doubling or tripling the yields expected in ordinary farming.

Because this simple method has the additional advantages of very low expense, and even greater increases in yields, it justly deserves consideration at a time like this, when conservation of the money people pay in taxes is as important as conservation of the soil. This method puts the farmer on his own to a large degree, and is even more desirable on that account.

This is but one of many ways in which it may be possible in the next few years to reduce the expenses of government. It may not be possible, as in this case, to find for each project new methods that are both better and cheaper than those proposed. It may, indeed, be wise in some fields to spend more than is now being spent; but it certainly will pay to investigate intelligently every item in each budget, to the end of getting the most possible in needed service for the money expended.

Ultimately, in all probability, we shall have surpluses of such magnitude as have never been dreamed of before. Several known factors tend in that direction, including the new erosion control suggestion just made. The surplus prospect creates the need for so redesigning our farming system that farmers will not hereafter *inevitably* grow more than they can market. One "sacred cow" of American agriculture is the supposedly necessary rotation of crops. Here is one way in which rotation of crops is justified. A definite acreage of hay land is needed. The tiny seeds from which grasses and clovers spring supposedly cannot be germinated successfully except in land which has been made relatively weed free by a year in corn, followed by a year in a small grain. The acreage that must be put to corn, and later to wheat, then, bears a definite relation to the acreage needed for hay. Since there have been in recent years progressive improvements in yield per acre of both corn and wheat, it should not be mysterious that we seem fated to produce corn and wheat surpluses. Increased dairying, incidentally, has served to relieve the hay surplus following substitution of tractors for horses.

Once this necessary relation between definite crop rotations and persistent surpluses has been recognized, we shall be able to glide smoothly into a system which might also be called a rotation but which has the advantage of removing the acreage-ratio requirement of the present rotation system. The burden of determining the acreage to be put to a given crop will again be placed where it belongs—on the law of supply and demand. No farmer will need to grow a crop for which there is not a normal demand, just so he can get his land back into some other crop he must have for home use.

This tendency, before mentioned, to grow bigger and bigger yields per acre is going to make necessary various changes in other ways, also. Fortunately, in every case where surplus production seems inevitable, there seems an advantageous change of plan which should make things easier for the farmer; and, happily, should cheapen the cost of production for him. This should be good news to non-farm people, for it means lower living costs for everybody, once the better farming system has been generally adopted. Less land, progressively, will be needed for the crops that heretofore have occupied most of our farm areas. There will be, therefore, acreage which may well be used for growing more of crops that have never been grown in sufficient quantity. Moreover, farmers will have more time to give to the culture of new and different crops. Foods that heretofore have been expensive luxury items may appear frequently on our tables, and at reasonable prices.

One crop that almost every state in the union grew

when its soils were new may again be generally grown, when the soil has again been renewed by the return to management methods that closely imitate nature. That crop is flax, now grown profitably in but a few of the states.

Grass, the "dictator" crop of the present crop rotation system, and a crop which should grow perennially wherever it has once got started, is likely to be a much more prominent part of the farm landscape of the future, for the reason that the proposed new cropping system will make it unnecessary to lavish farm manure on other crops, and the grassland can receive it all. Other improvements in grass culture are likely to be adopted in the coming years—so significant as to bring about corresponding improvement in our livestock industry. We are apt to learn much in future about methods of grass culture practiced in New Zealand and other grassland areas.

These are but hints as to possible revolutionary changes that can be effected in American farming—every one of which points to lower production costs for both crops and livestock. These changes will not be made, in all probability, unless many non-farm people become sufficiently well-informed to campaign intelligently for their adoption. As in few other fields of work, farm practices have become conventionalized. The customary practices are not merely habitual, but are justified by intricate, specious reasoning, and have to some extent become a sort of creed. The campaigner for new ways must be prepared to persevere in spite of opposition until success has crowned his efforts. This is one of the ways in which your country most needs your services in the immediate future.

8

SELF-DEFENSIVE FARMING

I F WE MAY CONCEIVE a new order for the United States in which an outworn psychology of high prices, surpluses, and traditionalism shall have been eliminated, it is quite possible to foresee an end to the kind of agriculture which has come more and more to rely upon government aid.

Agriculture is really the key to our whole future economics, but an institution that must depend for survival upon repeated transfusions from the public treasury scarcely deserves the role assigned to it. As long as this industry continues to require public sustenance, the honor of being the "key" to our *present* economics must go to the taxpayer, rather than to the farmer whom he sustains.

Fortunately, there are abundant means for realizing the new order. Under duress of reality, our psychology is being changed. In methods, we will adapt ourselves to two sound principles: (1) the retention of all the advantages that machinery has given us over the other agricultural populations of the world, and (2) the development of methods that will reduce costs and increase returns per acre.

Following the general policy of retaining our machinery advantage and adding an improved method of feeding our crops, we will proceed to redesign American agriculture on a plan that will be much less expensive per acre of crops grown—approximately one-half the present cost of growing both crops and animals.

The point of attack will be to immobilize the existing crop rotation by producing each crop hereafter on the field which best adapts that crop to the general farm scheme. For instance, haying is easiest when the fields are close to the barn. Pastures are far more convenient if they are in fields supplied by springs or streams. Silage corn can be most conveniently handled into the silo if the corn grows near the barn. There are few farms where the layout of fields is such that a given field can grow corn, wheat, or hay equally well from the standpoint of farm management. Rather, one of the difficulties of managing a crop rotation is apt to be getting in the hay from the back field in wet weather. If the hay can be grown permanently in the field or fields near the barn, this trouble will not recur. Incidentally, in the new scheme, the meadows and pastures may receive the entire supply of farm manure, for the other crops will be manured otherwise.

Corn, if this crop is needed, may follow a winter-grown crop of any kind, the latter being mixed into the soil surface as the land is being prepared for the corn crop. (Note that, since we shall no longer follow corn with a grain crop in order to fit the land for renewal of the meadow or pasture, there is no longer the necessity of planting a given acreage to corn for this purpose. This reduces the probability of corn surpluses by making it possible to grow as much or as little corn as is needed or can be marketed profitably.) Some difficulty with weeds is likely the first season. Weeds always grow better in a well manured soil—just as is the case with the crop you are trying to grow. You see, when a considerable amount of organic matter like green rye, or vetch, or other material has been mixed into the surface of the soil, the resulting soil is just as truly manured as if it had received a good application of manure from the barn. Thus, it suits weeds perfectly. In view of this fact, every effort to rid the soil of weeds ought to be made before the corn is planted. If possible, delay the corn planting enough to permit the killing of two or three batches of *germinating* weeds. This is done by disking or otherwise thoroughly stirring the soil surface two to three inches in depth every week to ten days. Then, when the corn has been planted only one or two cultivations should be needed; and these should be done while the crop is less than a foot high. Cultivation after the corn has got much higher will destroy many crop roots that will be feeding in the surface where the green manure is decaying.

By such methods as suggested, modified to suit conditions of the various crops and regions of the country, almost any crop ordinarily grown in rows for cultivation can be produced. The yield to be expected will depend upon the quantity of material mixed into the surface, the extent to which the soil has been depleted before the treatment, the amount of rainfall, etc. In many cases, probably not all, it will be found that the soil surface will take in all of any ordinary rain. This treatment of the soil is thus a definite check on erosion, because soil cannot be carried away by water erosion if there is no water to flow over its surface. By annual repetitions of this treatment—seeding a winter grown crop to precede the regular summer crop —the land may be made to produce yields that will increase each year over those of the previous season, until production finally reaches the top possible. This top may easily be several times as much as has ever been known to grow in the region since pioneer days.

Further unexpected effects will be noted. After a few seasons, in all probability, there will no longer be weeds to fight. This will happen, if it does, because your repeated cultivations have eliminated the plants as the seeds were germinating, so that no weed plants had a chance to mature seeds; and, therefore, no weeds could grow. In land that overflows this cannot be expected to happen, for each annual flood will bring in weed seeds from upstream. Also, birds and animals bring in seeds; and the seeds of some kinds of weeds are brought in by the wind. However, some men have reported that after they had managed their corn crops in this way for about three seasons they could scarcely find a weed. I have seen land that had been managed without plowing for ten years or more, and no weeds could be seen. This land, however, had been in crops broadcast solid over the surface—not row crops. The farmer may be surprised to find that he no longer has noticeable smut, European corn borer, or other disease or insect troubles. That has been my experience and the experience of others as well.

Wheat, which would follow the corn in rotation, will have its own field. As an annual manure crop for the wheat, any quick-growing summer crop is excellent. Sudan grass, soy beans, millet, rape, buckwheat, or any other easily seeded crop that grows off quickly in the community is all right. In Missouri, for more than twenty years farmers have been following a rotation that is exactly what is suggested here, except that they have in that state a crop for summer growth which reseeds itself each fall for the next spring. This makes it unnecessary to bother to seed the summer crop. This magically perfect crop for this purpose is lespedeza, which is adapted to a good deal of the country but it not known to grow as far north as the northern tier of states.

Because the idea is new and its application will increase the demand for seed of the customary summer green manure crops already mentioned, it may be difficult for a farmer to seed a crop to precede the wheat. In such a situation he should not overlook the possibility of using corn—of which he may happen to have a surplus. Corn is an old stand-by of dairymen. Besides its customary uses for grain and silage, it has often been employed as a catch crop in summer to relieve failing pastures. For this purpose it is seeded quite heavily, because of the extra size of the seed compared to the small grains. For keeping up milk flow in dry summer weather, the dairyman cuts each day enough of this tender sweet corn for his cows. Of course, in these days of insufficient help, little such extra work is done. Growing corn as a green manure for mixing into the soil, though, does not involve very much extra work; and its use as seed may make a green manure crop possible when otherwise there could be none.

One caution should be observed in so using corn. The growth should be mixed in when it is not more than two to two and one-half feet high. Otherwise, unless extra-heavy equipment is available for the work, mixing it in may prove impossible. If the stand is very thick, it may be safer to begin working it in at about eighteen inches in height. In case of wet, warm weather, corn makes astonishingly rapid growth; it is better to put it into the ground a little early than to run the risk of having it grow too high to mix in successfully.

Men whose land is clay will have serious doubts about the possibility of cutting into their soil at all with a disk harrow. Indeed, there is good reason for such doubts in the beginning. It may be necessary the first season to double plow. (Double plowing, in which the second plowing is a couple of inches deeper than the first, serves to open the way for the entry of mixing equipment such as the disk harrow. The two plowings should be done within a few weeks of each other, at most; and it must be expected that the second plowing will leave a surface that will be disgusting in appearance to farmers who have always taken pride in their neat furrows. There will be no neatness about it. It will be a mess. But the subsequent disking will improve its appearance greatly.) After this first season, no special treatment will be necessary, unless the land has been plowed meanwhile. If annual green manure crops are mixed into the surface, the soil surface will become highly granular, and will remain in that condition con-

tinuously. Clay that is granular will offer no more resist-
ance to disking than does sand. The trouble is all at the
beginning of the reformation of practices. Thereafter the
work is considerably easier than the customary plowing,
and far less power consuming.

In Missouri, where a large percentage of the farmers
grow their wheat in rotation with lespedeza, producing
wheat year after year on the same land, the lespedeza is
grazed by cattle, and produces more than one hundred
pounds of beef per acre annually in addition to the wheat.
The wheat yields by this system are somewhat higher than
the average of those obtained—once in a three- or four-
year rotation—by farmers who have not learned this better way. Thus, land farmed by an intensive green-manur-
ing system is proved by the Missouri practices to be considerably more productive. My guess is that, after a few
seasons of liberal green manuring without grazing, a
farmer may confidently look forward to producing on his
formerly indifferent soil wheat yields equal to the best
the Chinese can show. I base this guess on the assumption
that the repeated green manuring will prove to be the full
equivalent of the Chinese additions of human wastes. As
far as I know, nobody since the early pioneers has grown
one hundred bushels or more of wheat per acre in this
country; however, I know of no one who has treated his
land for several consecutive seasons to a liberal green manuring in preparation for the wheat crop. The possibilities
will remain in doubt until some of our farmers have followed these practices for a few seasons.

Rotation addicts will wonder what provision is to be

made for the renewal of grass stands. There are, indeed, serious questions that must be answered in this connection. Where there is plenty of land the answers are easy. Elsewhere the solution is far less simple. For the farmer whose space for hay and pasture is very closely limited, the best answer that can safely be made now is that he should follow strictly the practices recommended by the Agricultural Extension Service and the Soil Conservation Service. For some years these services have been showing farmers how to stimulate the growth of more and better grasses on land which is growing insignificant quantities of indifferent plants. The methods they recommend are highly successful, but are expensive in comparison with what can be done with a more liberal land area. Incidentally, farmers who wish to know a quick, simple, and effective method of starting alfalfa on waste land should write to the Ohio Experiment Station, Wooster, Ohio, asking for the Station's reprint on alfalfa culture from the *Bi-monthly Bulletin*, May-June, 1943. For farmers who have trouble starting alfalfa, this reprint will be invaluable.

If the yield increases suggested as possible for corn and wheat prove true to the "blueprint," even farmers who now have but limited acreage for hay and pasture will have to reduce greatly their grain areas. In consequence, much or all of this former grain acreage may well be put to grass for hay or pasture. For these and for others who have plenty of land, cheaper and equally effective ways of renewing grass stands are applicable.

The wise use of fence, preferably electric for convenience, may prove to be the key to pasture renewal once in two, three, or four years. In other words, if there is enough land to permit the practice, one-half the pasture may be allowed to grow ungrazed alternate seasons. This might be thought of as a natural, uncultivated green manuring process. The year of rest gives the grass leaves a chance to promote needed root extensions which are impossible under conditions of heavy or even medium grazing. Also, the decay of the fallen mat of the previous season's tall grass will supply the grass with a soil solution considerably richer in minerals. As a result, the grass will be more nutritious grazing for the animals after this season of resting. To farmers who have come to believe that no improvement in yield can come about except through the addition of something brought in from the outside, these suggested effects will seem impossible—the more so since there are few if any experimental figures to bear out the expectation of increased yields. I shan't trouble to argue the point. The way to conviction is trial. The truth will come out when once a farmer has actually allowed a pastured area (maybe just an acre or two the first season) to grow wild for a season. After he has proved my plan incorrect by trial will be the time for him to argue the point. (In all this, of course, the assumption is that there is actually a fair stand of ordinary pasture grasses on the land that is to be given a sabbatical year.)

Less effective, but a way of improving pastures somewhat, is the resting of one-third, one-fourth, etc. of the land. In all probability these less frequent grazing holidays would under many conditions serve to maintain productiveness at a high level. The more often the land can be given this chance to renew the stand the better will be the resulting improvement.

Unfortunately, there is one objection to the pasture renewal suggestions just made: there is always the fire hazard to think about. The plow's chief virtue, perhaps, was its effective removal from the surface of all fire hazards; for anything that will rot will burn. Conversely, then, the dead and highly decayable wild growth of pasture and meadow land is particularly to be watched as a possible fire hazard. Partly because of this dangerous aspect of pasture renewal, but more because it has some compensating merits, I hope to develop on my own farm within the next few seasons the pasture and meadow renewal practices used successfully in New Zealand, where most of the land is in grass. This method, if it proves as effective here as there, will renew the grass in from six to eight weeks, permitting grazing to be done on the entire area during part of every season. Thus, any dead, dry grass is less likely to be tall and heavy in growth.

My expectation of these methods is that they will permit natural processes to do for the farmer what his pocketbook has been trying unsuccessfully to do for him heretofore: fully maintain—even increase greatly in many instances—the productiveness of the land. Yet this will be but the first noticeable effect. Beyond high production will be higher nutritive quality than this generation has known, and virtually complete freedom of the grass from attack by insects or infection by disease. The higher nutritive quality in grass will, I hope, bring back the conditions which prevailed in this country when our land was young. If this happens, it will again be possible to market prime beef from the pasture direct. For the benefit of those readers who are farmers or stockmen, to whom this expectation will seem but the dream of an idealist, I quote an everyday practical farmer in one of our northeastern Ohio counties, where pasture today is generally poor: "In the early days, fat steers were sent directly from grass to market."[1] This farmer is the fourth generation of the same family on his farm and is now about ready to turn it over to the fifth. He should know what he is talking about. Because his description of the 140-year history of this farm seems otherwise in keeping with generally accepted ideas, I believe what he says about the ability of the lush grass to produce good beef in the early days must be accurate.

Protagonists of fertilizers, lime, etc. will be inclined to scout the possibility that the benefits they confer can be obtained without their use. There is no formal and official confirmation of the fact, of course; but we do know that even today, after unknown centuries of culture that removes annually much more mineral nutrients than our cropping removes, the peasant farmers of the world continue harvesting high yields of nutritious crops unaided by these supposedly imperative requirements. I see no reason for supposing the farmers of the United States are unable to do here what farmers everywhere else in the world do without even the advice of trained government specialists. The negative fact that no official confirmation

[1] Quoted from an article by Howard M. Call, "140 Years on Our Farm," *The Land*, Vol. IV, No. 4 (Spring, 1945), 143.

exists is to me far less important than the positive example
the farmers of the rest of the world have set for us.

This outline for the conditions easy-going future farmers may enjoy is a mere beginning of the multitude of new
practices that will emerge eventually. The importance of
this chapter consists in the fact that it shows definitely that
agriculture can produce food and fiber at cheaper rates
without reducing, even on the basis of lower selling price,
the income of the farmer. Since it is possible for agriculture to lead the way in the downward spiral of values our
country must negotiate, what are we waiting for?

As I noted earlier, the elimination of many customary
expenses from farming will of necessity mean serious loss
of sales for those whose products farmers will no longer
use. Realization of this fact has in the past prevented such
action as is now being proposed; and, without the information this chapter carries, the elimination of fertilizers,
lime, etc. (while men still plowed and farmed in a close-
knit rotation) would have resulted in serious reductions
in crop yields. Consequently, no one should be blamed
because action in this matter has been so long delayed.
Rather, let us get busy now and right ourselves agricul-
turally, to the end that we may also right ourselves indus-
trially and commercially.

9

FARMER MEETS CHEMURGIST

IF YOU HAVE NEVER MET a chemurgist, your education is far from complete, regardless of how extensive it may be. Chemurgists are a breed apart, and in a very real and laudable sense they lead double lives—almost to a man. Wheeler McMillen, for example, has presided over the national organization of chemurgists for several years, but in his business life he edits *Farm Journal and Farmer's Wife* of Philadelphia. Henry Ford was one of the very early chemurgists, and it is scarcely necessary to mention his "other" life. Dr. Karl T. Compton, president of Massachusetts Institute of Technology, is no less concerned about the development of new ways in which farm grown raw materials may be processed by industry to the profit of both farmer and industrialist. There is Louis J. Taber, too, past Master of the National Grange, farmer, and businessman. He is a vice president of the National Farm Chemurgic Council, representing agriculture. There are many others equally important in various professions, industries, research institutions, and in agriculture. The list includes most of the big names in every important activity that conceivably could use farm products in new ways.

If you wish complete information about this group, address a request to National Farm Chemurgic Council, 50 West Broad Tower, Columbus, Ohio.

Having thus introduced the agency which for more than ten years has been quietly doing something rather than wringing its hands about the farmer's problems, let us see what it has been doing. First, note what Dr. Compton had to say of the group's past and future in 1937: "To my mind, the most significant of all encouraging signs is the phenomenal growth of this farm chemurgic movement which is sweeping the country despite opposition from those who misunderstand it or who believe that their personal interests will be served by its failure. But it will not fail, because it is pointed in the direction of progress; it is based on the new philosophy of creating wealth and opportunity for all, rather than the age-old instinct of taking wealth from others; it is essentially co-operative between agriculture, industry, and the general public, rather than competitive between them." A list of the individual research projects in which these chemurgists are interested would take the rest of the space assigned to this chapter. We can mention but a few from a list of those in progress most recently.

The possibility of utilizing oil from tomato seeds (a factory waste product) is under investigation. Research on cotton for tire cords; cork substitutes from the pith and fibers of sugar cane, corn stalks, and peanut hulls; sirup from tangerines; rayon fiber as fine as silk; apple sirup from culls—all these are new developments. Various agencies are growing belladonna and caraway, some are producing industrial alcohol from wood waste. Penicillin production is being fostered and materially increased. Flax harvesting and processing machinery is being improved. The charcoal industry is being revived in Ohio. Noreplast, a new moulding compound developed from farm wastes, is already here. Fire retarding coatings, vitamin C from roses, bamboo research, cabbage juice found to be germicidal, home zero storage proved practical, improved methods of cattle branding to save leather and create a greater leather supply—these are taken from an incomplete list compiled by the Chemurgic Council and published in *The Chemurgic Digest* for January 15, 1945. Such is the partial range of this serious investigation.

In this same field, researches of Dr. Ernest Berl of the Carnegie Institute of Technology, Pittsburgh, have developed processes which make it possible for all kinds of organic wastes to be made into oils and gasoline. Dr. Berl believes that, if our supplies of petroleum should be exhausted, it would be possible to manufacture our entire peacetime requirements of oils and gasoline from farm-produced wastes. Indeed, he is confident that, if we now had to seek new sources for these necessities, the raw materials for their production could be grown on land that now is idle. He has in mind the thought that, because of the relatively high transportation cost of the bulky raw materials, the necessary processing plants would have to be small and numerous and set up at appropriate intervals throughout the country, rather than be located in the big metropolitan centers. Corn stalks and other light-weight materials can't be transported by pipe line.

The one special chemurgic development that has already become a regular part of farming over a wide area of the country is the production of soybeans. There are others of considerable importance, but the soybean has made the most spectacular entrance into the farmer's scheme of cropping. Starting from virtual obscurity as a farm crop a decade ago, soybeans have become one of the leading crops in several of the states; and the end seems not yet to be in sight. Rather, it looks as if the future of soybean utilization extends with increasing possibilities far into the blue. I have before me a list of the known ways in which soybeans or the plant that produces them can be used. It embraces no fewer than seventy-seven items; they range in character from fresh soy milk to rubber substitutes. He would be rash, indeed, who would attempt to predict any limits to the uses of this Far Eastern immigrant.

As shown in the previous chapter, summer-grown green-manure crops mixed in, preceding wheat, offer the big chance for greatly increasing the yield per acre of wheat. Soybeans may be the summer partner of wheat in that arrangement; and, in sections of the country where the season is long enough to mature the beans, the two crops per season—a crop of wheat and a crop of beans—may be harvested. In the South, where this might be done, wheat has not been a favorite crop. There the cowpea has been preferred to the soybean. In much of the Corn Belt, both wheat and soybeans occur in the cropping scheme, sometimes in the regular rotation, but not making a single one-year, two-crop rotation like the wheat-lespedeza rotation of Missouri.

One possibility to be watched for when farmers begin to precede all regularly harvested crops by a green-manure crop is that the time required for crops to mature will be shortened. This speeding up of maturity has been reported already, in some instances officially, as a result that follows the surface incorporation of green-manure crops in actual practice. To what extent the growth period of various crops can be shortened in this way is still a moot question, but time saving does occur, and for a very good reason. Weather is always a very important factor in producing crops, but no weather can make a crop grow rapidly if there is not also abundance of food material available and quickly transferable from soil to plant. When the soil surface is a teeming mass of decaying organic matter literally taken over by the roots of near-by plants, the stage is set for optimum weather to produce optimum growth. It is to be assumed that in our ordinary farming we seldom have the weather, the plant food, and the crop roots all teamed up and ready to rush business at the same time. If the wheat-soybean combination should result in lopping off a few weeks from the growth period of each of these crops, it is not unlikely that, over a considerable part of the country, these two crops could be established as partners in a rotation that would produce both a food and a chemurgic material which is also an excellent source of human food. This is not something to be counted upon; rather something to be hoped for and used to advantage if it proves possible.

Another chemurgic possibility that will bear looking into is flax. For a number of textile items, linen and cotton are direct competitors; and we can't afford to forget that we have a cotton surplus of the first magnitude. However, there is a tremendous field in which linen produced in this country could replace imports solely. You need but watch women buying imported linen articles as fast as they can be put on display—almost without regard for price—to suspect that American farmers may be missing a bet.

It would be useless to claim that fiber flax can be produced and disposed of as conveniently as, say, wheat or soybeans. These crops are simply seeded, harvested, and marketed, requiring relatively little total time in their production; moreover, the farmer rides a tractor most of the time he is growing these crops. Flax can be grown as easily, probably, but before it is marketable much dirty work that can't be done while one rides a tractor is necessary. One thing is sure: haying weather is not necessary for harvesting it, for its preparation for market can be done only after it has been "retted" (or rotted) by exposure to weather so that the fiber can be separated from the worthless portion of the stem. There is, therefore, a new routine to be learned, as well as provision to be made for handling the crop properly, before any farmer can be justified in deciding to go into flax. Yet, even this extra work might prove to be an advantage, for, as the plowless routines proposed in *Plowman's Folly* and referred to in this book are developed, the farmer will begin to find himself with more and more time on his hands, and to have a crop in production that will make this time profitable may be desirable.

The most that can be suggested at this point is that the production of flax and its preparation for market should be studied. One source of information on this subject is the National Farm Chemurgic Council, referred to earlier. For a small charge, non-members can obtain the latest information on developments in this field. There is under development new apparatus for retting flax, and all who are considering the production of this crop should know the latest developments.

The one fact about flax that is most pertinent to this discussion is its known adaptability to regions where wheat does well. Since in all probability some of the present wheat acreage will have to be abandoned for the sake of keeping production within market bounds, flax is a legitimate candidate for that acreage—if, after investigation, the farmer is willing to make the necessary arrangements for its production. It is thought that in pioneer days flax was grown in practically every state of the union. It probably requires for its best production soils that are better than our average soil is today. We are going to have those better soils as soon as we decide to give our crops a decent chance by letting them use the organic decay that lately has been wasted by our deep plowing. These improved soils probably will produce excellent flax, if we make up our minds to try it.

Like cotton, soybeans, and peanuts, flax is a double-duty crop. It produces both fiber and oil—linseed oil, inedible but in demand in paint manufacture. Generally speaking, flax grown for the production of seed is not used in fiber production, and vice versa. Details such as these must be learned by the prospective grower. If he wishes to grow flax for its seed, disregarding the fiber value, the process is simpler and may be little more complex than the production of the usual crops. At any rate, flax for fiber or seed deserves consideration by farmers who need to reorganize their farming for more efficient production.

There is also the sweet potato that is grown to some extent in most of the states. It has gone chemurgic to a certain extent. How extensively this crop will be used in the future for starch production and other industrial uses depends upon the cheapness with which it can be produced, the market demand for the roots as food, and other equally practical considerations. If it should happen that the sweet potato crop of the South is in such demand for starch as to affect the supply shipped to the North for food, there will be an opening for northern farmers to become growers on a small scale. Also, it has been found that dehydrated sweet potatoes make excellent feed for livestock. The northern farmer may find a market for his production of this crop in the dry form; or he may make use of sweet potatoes to some extent as feed in place of corn. Farmers with time on their hands may well look into the requirements for dehydrating this and other crops that do not store easily.

The successful production of sweet potatoes is made far easier if weeds are not a serious factor. Farmers who have gone through a few seasons of plowless operations and find that their land no longer produces weeds in appreciable numbers may well consider seriously the production of a few hundred bushels of sweet potatoes annually for whatever use offers the best return. Dairy cows are extremely fond of them, and so are hogs. If they have not cost too much to produce, they may also prove profitable if sold to a chemurgic processor.

It is generally agreed that, for the future, it will be well for the United States to be self-sufficient in rubber, as well as oils, cordage, and many other things that become embarrassingly short in wartime. This necessity opens a large field for farmers who will in the future have land they cannot use profitably in growing the customary crops of today. The richer, less weedy soil we shall have in the future assures space for inexpensive production of these strategic necessities. The equally strategic importance of producing these crops at the lowest possible cost should be obvious. Otherwise, unless the government protects home production by high tariff walls, importations will come in, to the ruin of the home industry. Low-cost production surely is better economy than protective tariffs. When we have developed our land again into spontaneously rich, practically weed-free condition, we shall pit our machinery advantage against the low-living-standards advantage of competing nations; and there is no reason why we should not win out easily. If we coast along as we have been doing for the past generation or two, hugging the starry-eyed dream of our vanished grandeur, we are certain to lose out in the unequal competition provided by low-cost foreign production. We can, if we will, kill off that competition by the only sound economic weapon known—low costs.

Then, there is the entire field of plastics. American industry would gladly make plastics from appropriate farm-grown materials if they could be bought on a competitive basis. For more than ten years chemurgic authorities have been stressing this field as one of the means by which farmers can add to their dwindling incomes. Farmers, however, caught in the dilemma of high production costs supposedly essential, have been unable to take advantage of this opportunity. In all sincerity, professional agriculture has been advising farmers along lines which have necessarily kept production costs high. And the entire realm of theory in farm practice has been perfectly logical, if the correctness of plowing is assumed. Moreover, professional agriculture, as truly as farmers themselves, has believed in the rightness of plowing. It was to be expected that procedures logically consistent with this point of view would be recommended. The train of high costs which has followed also becomes logical.

These are some of the possibilities to which farmers may look forward for replacement, at least in part, of income they stand to lose from the cheapening of the per-unit cost of their crops and livestock. As cost reduction is achieved in business and industry, following agriculture's lead, it will no longer be necessary for the farmer to have as large income in dollars in order to have equally good purchasing power. The picture as a whole, then, should be much improved.

It is to be hoped that those whose business it is to advise farmers will promptly orient themselves to the necessary economics of our situation and revise their agenda in keeping with the realities of that economic compulsion. Too long have we of the United States been a special case in economics. As soon as we have rid ourselves of our unessential costs in agriculture as well as in other fields, we shall be able to deal with our world neighbors on reasonably equal terms, to the advantage of all parties.

10

KINDLING THE NEW IDEA

SETTING THE AMERICAN FARMER once again upon his own two feet under conditions that will enable him to manage for himself as the rest of us must do is, I believe, close to the key to a solution of the entire American economic problem.

This statement is not intended as an indictment, or even a disparagement, of efforts that have been made hitherto to relieve the farmer. Indeed, it is doubtful if our best informed economists could determine definitely the extent to which the general situation has been affected, for good or for evil, by such regulatory work as has already been done. Some evidence suggests that regulations have caused "black market" operations; other equally trustworthy factual material as clearly indicates that the farm situation itself may have been salvaged from ruin. Take your choice.

Fortunately, we need not approve or condemn past efforts in order to launch now the regime that should normalize agriculture and tend to solve many non-agricultural problems as well. What we need and must have for this purpose is the earnest co-operation of everybody who is concerned with our agricultural establishment—to the end that the necessary changes may be effected in farm policies and practices. No war-bond drive or other campaign for the general good has ever been more important, and none more necessary to the welfare of the country as a whole, than this proposed campaign to initiate at once a more realistic American agriculture. It can be accomplished immediately, assuming, of course, the necessary correlation of effort on the part of everybody concerned. The change is, in fact, already under way; but it will gain momentum slowly—as the idea kindles from man to man —unless conscious effort is made to realize the benefits of the idea at once.

This problem—our sick agriculture—must be understood before we attempt to apply the remedy. As the physician diagnoses the ailment from its symptoms, we must search out the causes of agriculture's present impotence in order to deal intelligently with the problem. We might well review the case history.

If you are unacquainted with farming, you will wonder how the agriculture of this progressive country could now, of all times, be in the doldrums. There is every reason for the uninitiated to expect that American agriculture should be leading the country's progress, instead of being a ward of the government.

You are already aware that during two centuries or more this land was being cleared of forest, or of the age-old sod that covered its surface. As rapidly as the land was cleared or the sod broken, crops were grown, and the surplus found its way into world trade. Incidental to this disposition of our surplus crops, we accumulated here the greatest quantity of gold ever brought together in one place. Everybody in the country shared in the resulting wealth. This was the one country in which everybody could have all the money he needed. Day laborers here lived almost as well as the royalty of some countries. We had a near Utopia—without knowing how to keep it.

As farmers cleared their land for cropping, they soon found themselves unable to manage by hand methods all the land at their disposal. There soon developed a demand for machinery to operate the broad acres. Inventive men were alert to supply this need, and before long there were busy factories turning out mowers, reapers, binders, and other implements that made it possible for every farmer to manage several times as much land as he could have done otherwise.

These early farmers have been blamed for the ruthless manner in which they mistreated their soil. I believe we may profitably consider what we would have done had we been in their situation. Picture, if you can, the farmer who had a hundred acres under fence, with several hundred more acres to be cleared. Without machinery he could not begin to produce and harvest crops from his cleared land. With machinery he could participate in the incoming stream of gold to the extent his crops justified. To him, soil was commonplace; machinery was precious. To you, if you had been in his place—as your great-grandfather probably was—the relationships would have been precisely the same. We should not waste time, therefore, blaming the men who wore out one farm and moved to another, even though, in the light of our present knowledge, they were doing a very unfortunate thing.

The use of machinery was almost a self-defensive measure for these farmers. Their soil was so rich that if they left it untilled for a year or two it began to look like a young forest again. They had to have machinery to prevent their land from lapsing back into woods. It is easy to understand how the very richness of the land forced measures which to us seem poor management of the soil. In those days, we should remember, land that today may be a pure red, yellow, or gray tint was pure black. There was no possibility of classifying soils by types, as is done today; the distinguishing characteristics were completely masked in the black of decaying organic matter. These are facts we are apt to forget in our zeal to incriminate the men responsible for the first erosion of our soil.

To continue our efforts to locate the underlying causes of the present situation, let us try to understand why such scientific findings as were available to these early farmers were not more extensively used. For many decades, experiment stations have been investigating every farm problem which seemed to them to deserve study. From time to time reports have been made of the results of these studies. Recommendations to farmers, oftener than not, have been ignored, in many cases without the slightest flicker of interest. The passive rejection by farmers of scientific information designed for their benefit seems uncalled for, but I believe a further study of farmer psychology throughout our past will show that such behavior is strictly in the American tradition.

Think of it realistically! You would never expect the owner of a quarter-section of unmortgaged land who has cash in the bank and plenty of machinery to think of economic matters in the way a peasant farmer must think of them. Peasant farmers of the rest of the world have done a better job of managing their land, traditionally, than the American farmer has. But think how different are the viewpoints of the two. American farmer and peasant can't possibly think alike about the two items, machinery, and soil. The American has perhaps thirty times as much land; possibly a smaller family to help operate the farm. The peasant must save every vestige of organic matter, compost it, and use it to the best possible advantage. To the American, neither compost nor manure means much in his life. Because of his relative wealth, his familiarity with machinery, and his relatively few helpers, he can't afford even to haul the manure from his barns. Instead, he buys commercial fertilizers and applies them in a fraction of the time that would be required to handle the equivalent in manure. No scientist advised him that the commercial fertilizers were better. He listened to no scientist, as a matter of fact; but if he did, he still would have bought and used the fertilizer. Why? Speed and convenience.

When we consider the conditions under which American farming has developed, the sins we have charged against the men who cleared the land and established that first agricultural "beachhead" probably have been overemphasized. Our incriminations have been largely misplaced. True, our ancestors were interested more in the newest farm machine designs than they were in saving their soil; yet we must remember that they had as much trouble keeping ahead of the encroaching forest as we have keeping out of the clutches of the sheriff. Efficient farm machines helped a great deal. Reading a dry harangue on the need for preserving his soil probably wouldn't have helped a bit. A realistic point of view is necessary in appraising fairly the pioneer's work.

We know that Washington noted here and there a spot where his land eroded; Jefferson, too, has been quoted as reporting contour farming on part of his land. However, until much later trouble from erosion was noticeable chiefly as recurring sandbars in streams. Everywhere during rains the streams ran red or yellow, depending upon the color of the native stone; but this all but universal condition probably was overlooked, because it was present everywhere. Nobody traced this stream coloring back to its cause. Hence nobody realized the impending danger in the gradual fading out of the dark smudge that characterized all of the land in the beginning.

In appraising the way both farmers and scientists looked at the land years ago, we must know the broad general scheme of our phenomenal farming system. The American farmer was an expansive fellow. He had to be. He worked the land "in the large" and thought of world trade in the concrete terms of his own crops. Seldom was he a detail man. Scientists who served him perforce gave most attention to investigations that would serve him best for the moment. There was little point to following a line of investigation the results of which would not be given consideration after they had been obtained. We might see fit to criticize scientific men for their failure to keep farmers up to date on the subject of erosion dangers; yet even that criticism would be ill-advised.

We must keep in mind always the contemporary circumstances that determined an action or a failure to act. Following through the earliest period of our farm history, we recall that most of the land was rich and required no fertilizer; also, there was enough unused land already cleared in many communities to permit a transfer of operations to this fresher land until the worn land had made a natural recovery. Such facts masked the fact, now realized but not thought of then, that gradually the land's ability to produce became a bit feebler with each passing decade. Successive generations of occupants of the land have found it necessary to add increasing amounts of fertilizer in order to maintain production at a profitable level. Use of fertilizers today is at the highest point in history, and current recommendations call for still bigger applications.

Throughout most of the period we have been describing, the attention of both farmers and their scientific advisers was directed to the quantity of crop produced, rather than to the condition of the land that grew the crops. Both custom and theory had established the use of fertilizers, for they did, in fact, increase yields. Thus the "necessity" for their use easily became a theory, bolstered by the familiar bank-account analogy. If you wonder how it could be that scientists themselves should have failed to note the contradictions of the universal landscape, please remember that throughout the earlier period they had been compelled to cater to the farmer's willingness (and ability) to spend money in order to save time or labor. That attitude is the key to much that has happened in the past century of scientific investigation. If an investigator had confined his attention to methods of retaining organic matter in the soil, his reward would have consisted in his being completely ignored by the very clientele he was supposed to serve.

Unfortunately, scientists themselves die and must be replaced. During the past century of scientific investigations in agriculture, there have been as many generations of scientific agriculturists as of farmers. The scientific traditions of a century ago (or fifty years ago, for that matter) are unknown to today's scientists. For proof, examine any agricultural textbook dating from prior to 1910. These older books discuss farm problems from a point of view that recognizes natural factors to a much greater extent than is the case today. Crops produced by the best soil, for instance, were known to be troubled little or not at all by insects and diseases. Today's scientists still agree that this is true, but they honestly (and let me emphasize the correctness of the adverb *honestly*) believe that the difference is not enough to warrant a farmer's action in omitting to spray or dust his crops as insurance against damage.

The real, though unrealized, reason for this conviction on the part of scientific men is the fact that few of them have ever seen a soil that was literally alive with vigorous decay. Such a soil, even if unfertilized, does grow crop yields as high as the highest science can achieve with all its effort. And crops so grown suffer no serious depredations from insects or diseases. The customary destructive insects may be present in small numbers, but little or no damage will be done. Often it is impossible to find the slightest evidence of disease attack. Such unusual evidence has not been observed by most scientific men, because their experience has all been with plowed land, which supports little decay at or near the surface. Hence it is easy to understand why scientists today are incredulous when they are told that such crop immunity may be achieved simply by adopting correct growing conditions.

From the above discussion it should be clear that our farm practices have grown up around practical rather than scientific considerations, despite the popular notion that American agriculture is the most scientific in the world. Its science has been mechanical or electrical, chiefly. Chemically, physically, and biologically, science has had to bow to convenience and speed. It could not have been otherwise during the earliest part of our farm history; and as generation succeeds generation, both farmers and their advisers lost the traditional clues that would have led them toward correct soil management practices.

It is not strange at all that as venerable a practice as plowing should never have been subjected to investigation during the period when convenience and speed had to determine the choice of practices. Farmers had to plow. Now, since the shadow of suspicion rests upon that ancient implement, agricultural scientists can turn to no trustworthy comparative evidence, except that of the last few seasons. Having omitted to experiment with crops on worn land that had been prepared by intimately mixing in bulky green-manure crops, as suggested in *Plowman's Folly*, scientists have no way to know what kind of results to expect. Farmers, too, are at a loss to appreciate the possible improvements in yields, except as they have had some comparable experience themselves.

Almost every farmer has had some experience with plowless methods, but few have thought that the remarkable yield achieved might have resulted solely from the better tillage methods. Dozens of farmers, when the underlying reasons have been explained to them, have told me about the greatly improved crops they got from the back forty the year they didn't have time to plow for wheat. But they hadn't been impressed enough then to ask why it happened.

Knowing the historical facts that have contributed to our present agricultural plight, we can understand why neither scientific men nor farmers would accept easily the ideas of *Plowman's Folly*. The soil conditions anticipated by that book have never been attempted by farmers —for practical reasons. When a large quantity of organic matter was to be disposed of, disking was slower, or seemed to be slower, than plowing and apt to leave a trashy surface. That was reason enough for farmers or scientists of practical bent. This accounts for the perfectly logical answer a surprised news service man got last year when he inquired of a state agricultural extension office as to the reception *Plowman's Folly* had been given by farmers. The director told him that real farmers of the state were paying no attention to the book; which is precisely what should be expected in a year when shortages of both labor and machinery dictated that speed and convenience rule as never before. Farmers could not have done otherwise under the circumstances.

Has it ever occurred to you that the farmer's allegiance to speed and convenience is of personal concern to you? Increases in food costs have been built up item by item chiefly because of the desire for speed or convenience. Relying upon his economic and speed advantage acquired through machinery, the American farmer has assumed extra costs unhesitatingly. He has finally overplayed his mechanical advantage, which in recent years has lost its chief attraction as soils have declined. Accumulated costs today far overbalance the mechanical advantage, and competition from foreign countries (formerly our customers) can now invade the home market, because foreign agricultural costs are lower. Tariff walls tall enough to protect our farmers undoubtedly will be erected, but the thought isn't a comfortable one, even so. The American farmer would like to control world markets, even as his great grandfather used to do.

It is too bad that farmers did not learn long ago the wonderful advantages of plenty of organic matter mixed into the soil surface. Organic matter so used does everything that is needed for the correction of soil troubles. It replaces both lime and fertilizers, and does it so completely that, once it has been given a proper trial, farmers will wonder how they could ever have been so mistaken in their ideas about the soil. The evidence is to be seen in every fertile spot of the natural landscape; or, if that must be mistrusted, we may find abundance of similar evidence among peasant farmers everywhere. They must live from the land, as we must; yet they manage very well without the necessity for buying anything to sweeten their soils or feed their crops. Knowing this, it is regrettable that American farmers should have permitted themselves to roll up an enormous bill for supposedly essential costs, thus eliminating themselves from the world markets their forefathers enjoyed.

As has been shown, the entire proceeding developed so naturally that, in our geographic isolation, we had no chance to check our "progress" by the experiences of agriculturists in other countries. Indeed, as tourists we looked down our noses in unmistakable disgust at the small-time operations of the peasants we saw working their fields in Europe or elsewhere. All the while, until within the past decade, our average crop yields were declining sharply without the facts' being generally known. In more recent years, at great additional expense met partly by government assistance, average yields began gradually to improve. For the first time in many years, corn and wheat again began to be produced regularly in quantities that the domestic market would not take at the price. Then we learned the worst: our mounting costs had made our grain so expensive that the surplus could no longer sell abroad, as had been possible before our farmers developed the big bulge in their production costs.

In view of the otherwise fantastic economics of our country, the high cost of growing farm produce has never seemed strange; indeed, it has been of a piece with the rest of our scale of values. The system grew, weed-like and unplanned, out of the fabulous hoard of cash we piled up in the early days. A wealthy position which has disturbed our foreign economic relations ever since we first felt its domestic effects has blocked us at every turn. We do not control our costs, and for that very reason we are poor competitors in the world's markets. Our poorer world neighbors have fenced us in perhaps more than we know, thereby producing in us an isolationism that has little relation to practical politics.

Whatever our misconceptions in the past of proper relations to maintain with our world neighbors, prudence suggests the necessity for quickly adjusting ourselves to a point of view more acceptable to the rest of the world. Prerequisite to a permanently peaceful world, authorities agree, are satisfactory economic relations. There can be no political harmony without them. Large economic differences block world trading, which normally grows directly out of harmonious economies. If only for the sake of peace, then, we must set our economic house in order.

These considerations leave us no alternative but to bring about the greatest possible deflation of our farm production costs as the initial step in a general program of price and cost reductions. We can grow many farm crops at considerably less than half their present production costs. Up to the point of harvesting this should be true for most crops. The full possible reduction may not be reached in the initial approach to the problem, but the future already promises to bring very much lower food costs. It is important to remember, too, that this objective can be reached without in the least reducing the farmer's profit.

There will be a reduction in his total income, not in his profit. His net income may be greater eventually.

Following the farmer's lead, labor, convinced that for once no opponent holds a trump card up his sleeve, may gracefully accept lower wages, because of important savings in the matter of food costs. Net income, in the case of labor, need be no less. Indeed, it may well be far more in the end, for there are yet other generative forces.

Employers of labor, who must charge labor costs, along with other essential items, into the selling price of every spool of thread and every locomotive, will be able to write down the selling prices of their products—again without reducing their profits.

Food buying will be stimulated by the lower food prices exactly as it was stimulated, in spite of higher prices, by increased incomes during the war. This increased demand for food (because a dollar goes further) will be just as effective as if there were more dollars to spend for this commodity.

Soon, too, the manufacturer will find that, since he has priced his goods on a lower level, demand for them has greatly increased. He will thus be compelled to increase production, which will require more men or more hours per man. In the end, therefore, while everybody may handle considerably less money, there will not necessarily be less enjoyment of life. There is a strong probability that the multitude of ways in which production, transportation, and sales will be stimulated will bring about bigger business opportunities than we have ever known before.

It will be difficult, if not impossible, for people who have become adjusted to our present inflated scale of values to believe that results of the kind described here could follow general cuts in production costs. So implicit is our faith in the necessity for high scales of value that this discussion will appear to be pure fancy. Many readers are too young to recall this country's most famous cost-cutting experiment. For their information it may be well to recount it here.

Thirty years ago Henry Ford conceived the idea that the only way to make the automobile serve the largest possible number of people was to sell it at a price within the reach of every man who held a reasonably good job. In order to inaugurate the new policy, he demanded his raw materials and parts on the lowest possible cost basis. Before long, sales of Fords reached previously undreamed of figures. By further reducing production costs and selling prices, the Ford Motor Company was able to sell all the cars its assembly plants could produce—at one time as high as seven thousand daily. In justice to buyers who had purchased before the lower costs had been reached, the company made the price reductions retroactive and mailed refund checks to these buyers, thus equalizing the difference in purchase prices. This gesture evidenced a new kind of justice and was the finest of good will advertising. Sales increased until more Fords were sold than all other makes of cars combined.

Competitors and business leaders of the country were dumbfounded by such unusual ideas. They were sure Ford's venture would fail. Yet the over-all effect on the country was electrifying. Other manufacturers began to follow suit, and the automobile, as a result, became each year a little lower in price, in spite of improvements in quality. Whichever car you drove, the effects of Ford's original cost-cutting venture influenced both its cost and its value to you.

The suggestions of this book are in the same economic pattern. Presumably they have the same theoretical chance to succeed. They are different in that no one man can initiate them; they must be organized on a nation-wide basis, and that will take time. The lead should originate in Washington, making use of our already well-seasoned nucleus for the necessary national working force. Occasional farmers and others throughout the nation are working with the soil procedures involved. They will help. We need but the leadership.

Intelligently planned, deliberate deflation of farm costs, coupled with coincidental deflation of other value levels, should be possible to achieve without major economic displacements, provided enough appropriate educational effort precedes the action. Most of our adult population will need this preliminary coaching, and it must be as comprehensive as our war bond campaigns were. Without preliminaries to prevent fright, a deliberate deflation campaign might precipitate the worst economic crisis this country has ever seen.

Many economists fear, I suspect, that such a crisis is inevitable anyhow, regardless of what we do or fail to do. This fear, though never expressed, seems implicit in much of the public discussion that trots out all the premises of our postwar situation without trying to draw conclusions from the premises.

Never before in my recollection have economists been so uniformly agreed about anything as they seem to be now about the framework of conditions necessary to success in peacetime. All agree that our surpluses will not lend themselves to competition in world markets because our costs of production are too high. None suggests the obvious conclusion: that drastic reductions in production costs must be effected. Instead, some observe timidly that the government may be forced to shore up sagging farmer-income, or sagging industrial employment, in order to keep production going at present levels of value. This temporizing on the part of economists offered the opportunity for presenting the discussions contained in this book.

The necessary action extends to the whole field of human activity in the United States—a scope which, obviously, no book could encompass, even if the author were capable of so catholic an undertaking. No attempt has been made to go beyond the merest outline of the problem. There are men in various government agencies capable of the necessary elaboration of plan and method. To them we must look for guidance in the working out of the desired program.

From them we require wisdom, from ourselves the courage to try a new and better way.

uneasy money

HAS BEEN SET ON THE LINOTYPE IN THE

TWELVE POINT SIZE OF BODONI BOOK

AND HAS BEEN PRINTED ON

WOVE ANTIQUE PAPER

UNIVERSITY OF OKLAHOMA PRESS

NORMAN